REINFORCEMENT OF POLYMER NANO-COMPOSITES

Reinforced rubber allows the production of passenger car tires with improved rolling resistance and wet grip. This book provides in-depth coverage of the physics behind elastomer reinforcement, with a particular focus on the modification of polymer properties using active fillers such as carbon black and silica. The authors build a firm theoretical base through a detailed discussion of the physics of polymer chains and matrices before moving on to describe reinforcing fillers and their applications in the improvement of the mechanical properties of high-performance rubber materials. Reinforcement is explored on all relevant length scales, from molecular to macroscopic, using a variety of methods ranging from statistical physics and computer simulations to experimental techniques. Presenting numerous technological applications of reinforcement in rubber such as tire tread compounds, this book is ideal for academic researchers and professionals working in polymer science.

T. A. VILGIS is Professor of Theoretical Physics at the University of Mainz and a researcher at the Max Planck Institute for Polymer Research. He is a member of several scientific societies including the German Physical Society, EPS, and APS. He has written more than 250 scientific papers, three popular science books and two scientific cookbooks.

G. HEINRICH is Professor of Polymer Materials at Technische Universität Dresden and is also Director of the Institute of Polymer Materials within the Leibniz Institute of Polymer Research. He has written or contributed to over 250 scientific papers and book chapters on polymer science.

M. KLÜPPEL is a Lecturer in Polymer Materials at Leibniz University, Hannover and Head of the Department of Material Concepts and Modelling at the German Institute of Rubber Technology (DIK). He has published more than 150 scientific papers and is a member of the German Physical Society, the German Rubber Society, and the Rubber Division of ACS.

REINFORCEMENT OF POLYMER NANO-COMPOSITES

Theory, Experiments and Applications

T. A. VILGIS

Max-Planck-Institut für Polymerforschung, Mainz

G. HEINRICH

Leibniz-Institut für Polymerforschung, Dresden

M. KLÜPPEL

Deutsches Institut für Kautschuktechnologie, Hannover

CAMBRIDGE
UNIVERSITY PRESS

CAMBRIDGE UNIVERSITY PRESS
Cambridge, New York, Melbourne, Madrid, Cape Town, Singapore, São Paulo,
Delhi, Dubai, Tokyo

Cambridge University Press
The Edinburgh Building, Cambridge CB2 8RU, UK

Published in the United States of America by Cambridge University Press, New York

www.cambridge.org
Information on this title: www.cambridge.org/9780521874809

First published 2009

Printed in the United Kingdom at the University Press, Cambridge

A catalog record for this publication is available from the British Library

Library of Congress Cataloging-in-Publication Data

Vilgis, T. A. (Thomas A.)
Reinforcement of polymer nano-composites / T.A. Vilgis, G. Heinrich, M. Klüppel.
p. cm.
Includes bibliographical references.
ISBN 978-0-521-87480-9 (hardback)
1. Reinforced plastics. 2. Nanostructured materials–Inclusions. 3. Composite materials.
4. Rubber–Reinforcement. I. Heinrich, G. (Gert) II. Klüppel, M. III. Title.
TA455.P55V54 2009
668.4′94–dc22

2009013054

ISBN 978-0-521-87480-9 hardback

Contents

Preface

Why a new book about the science of an apparently old material? This question can be easily posed, when reading the title of this book. Indeed, filled rubbers are well known and well used in daily life. However, it is less known that recipes and the corresponding processing cycles of carbon black or silica filled rubber are extremely complex, which leads to a complex structure of the material in a wide range of length scales. Rubbers are classes of relatively soft materials without which modern technology would be unthinkable, similar to the case of metals, fibres, plastics, glass, etc. No matter where these rubber materials find their application, especially in tires and in a great variety of industrial and consumer products, e.g. motor mounts, fuel hoses, heavy conveyor belts, profiles, etc., the applications make high demands on rubber materials. The requirements are manifold, e.g. high elastic behavior even at large deformation, tailored damping properties during periodic deformations, great toughness under static or dynamic stresses, high abrasion resistance, impermeability to air and water, in many cases a high resistance to swelling in solvents, little damage, and long life.

Their importance for applied sciences and engineering is unquestionable, so why not collect the ideas and facts about these materials in a book? Aren't there many theories and facts around which many could form the basis for a review book? This would be, however, too simple, at least for us and for the completely different backgrounds of the three authors. Providing such a book is probably useless and not very exciting. Moreover, most of the theories that are around seem to suffer from too much phenomenology, too much diversity, and too much empiric reasoning.

Rubbers are far more than boring materials, at least from a theorist's point of view, at least from an experimentalist's point of view, at least from an engineer's point of view. Last but not least, from the materials point of view, simply because the function and the wide-ranging properties of the material depend on large variety of lengths and time scales. Filled elastomers are a typical example, where multiscale science plays a major role in the structure–property relationship. Imagine a car driver who needs to brake suddenly to stop at a very short distance. Can he, at the same time, imagine that this macroscopic, highly nonlinear process can be drawn back to certain and well-desired physical properties of the nanoscale polymer layer formed around the filler particles that are embedded within the rubber matrix? Can the car driver imagine the role of the filler network formed by the aggregated filler particles that form a random (cluster–cluster) percolating network? Or, how is the wet grip of the tire related to certain time and length scales within the tread rubber material that is excited periodically during sliding over a rough, even fractal, road surface?

The present book cannot give all the answers to all the questions, but we try here to develop a picture for filled elastomers, which joins basic theoretical ideas with practical applications. The basic ansatz here is therefore different. Starting from theories, we try to understand many, so far, empirical laws to provide more physical insight. We try to join different ideas together by using solid models. These, very often fundamental starting points will nevertheless lead to new ideas, new pictures, and new models. This is, what we, the three authors have done over the past 10 years in our common research starting from our three individual backgrounds. Thus the book has a very personal point of view. It is based on our own reach and based on the different attitudes of all three of us. It joins basic polymer physics, sometimes hard core theory, with experiments and at various places questions located in applications and engineering. This book is an attempt to provide more physical insight into the properties of materials, and therefore we try to relate most of the macroscopic features, which define the properties, to elementary physical pictures and models. To do so, we need a large variety of theoretical and experimental approaches, since a broad spectrum of lengths and time scales need to be taken into account. For us it was sometimes exciting to realize how purely theoretical results from simple models, e.g. universal exponents for frequency dependence of relaxing localized chains, transport themselves into measurable quantities, e.g. the relaxation time spectrum ruling the frequency dependence of the modulus, in certain time scales. Perhaps the reader can share our excitement here and there in this book.

Therefore, this book is indeed a kind of review book, but of our own work and from our own points of views. This remark needs to be understood as an apology to many other authors who will not find themselves quoted here, but also as an invitation to follow different ideas and different viewpoints about a classical material. If the reader is following this invitation, he can then perhaps agree with us. Filled elastomers are indeed classical materials, but they offer still many open questions and many possibilities for fundamental studies. On the other hand, cognition of our studies has been used by the authors to develop and to design certain kinds of future rubber materials based on concepts of rubber nanocomposite technologies.

In particular, this can serve as a tool for developing a new tire generation with improved rolling resistance, wet traction and wear properties, and in this way, break through the magic triangle of tire technology. However, this will not form part of this book.

T. A. Vilgis,
G. Heinrich,
M. Klüppel,
Mainz, Dresden, Hannover,
November 2008

Acknowledgement

The authors thank the German Rubber Society, the German Ministry of Science (BMBF), and the German Science Foundation (DFG) for support at various stages of the work reported here. The authors are in debt to Katja Tampe, Marina Grenzer, and Sven Richter for their critical reading of the manuscript and valuable technical help.

With special acknowledgment to Distinguished Research Professor Jim Mark, University of Cincinnati, acting as Polymer Advisor on behalf of Cambridge University Press.

1

Introduction

The reinforcement of composite materials is far from being a simple problem [1]. Reinforced elastomers, which find application in the car tire industry, are typical and well-known examples of that. Indeed, these materials allow a physical formulation of most of the problems and offer a suggestion for a solution. Complications arise due to the many length and time scales involved and this is one of the issues which will be examined in this book.

The basic aim of filling relatively soft networks, i. e. cross-linked polymer chains, is to achieve a significant reinforcement of the mechanical properties. For this purpose, active fillers like carbon black or silica are of special practical interest as they lead to a stronger modification of the elastic properties of the rubber than adding just hard randomly dispersed particles. The additional reinforcement is essentially caused by the complex structure of the active fillers (see, e.g., [2] and references therein).

The main aim of the present work is to gain further insight into this relationship between disordered filler structure and the reinforcement of elastomers. As a filler type we have chiefly in mind carbon black, which shows "universal" (i. e. carbon-black-type-independent) structural features on different length scales, see Fig. 1.1: carbon black consists of spherical particles with a rough and energetically disordered surface [3, 4]. They form rigid aggregates of about 100 nm across with a fractal structure. Agglomeration of the aggregates on a larger scale leads to the formation of filler clusters and even a filler network at high enough carbon black concentrations. Reinforcement is thus a multiscale problem.

These universal features are reflected in corresponding universal properties of the filled system. For example, the geometry and activity of the filler surface play major roles in the polymer–filler interaction: the physical and chemical binding of polymers to the filler surfaces depends on the amount of surface disorder. Aggregate structure is expected to be dominant at intermediate length scales and agglomerate structure at large length scales. Interesting phenomena like enhanced hydrodynamic

1

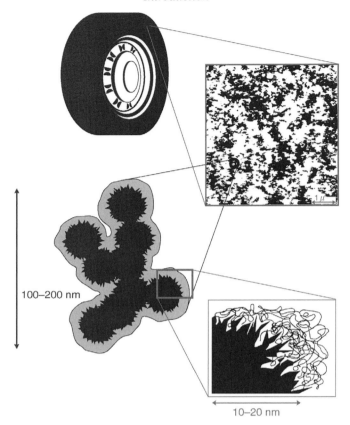

Fig. 1.1. Structural properties and scales in carbon-black-filled elastomers on different length scales.

reinforcement and the Payne effect can be attributed to the fractal nature of the filler structure. From these considerations it is clear that classical approaches to rubber elasticity are not sufficient to describe the physics of such systems. Instead, different theoretical methods have to be employed to deal with the various interactions and, consequently, reinforcing mechanisms on different length scales. Moreover, we have to indicate physical length scales as well. Considerable reinforcement can only be achieved if the length scales of the filler and the polymer matrix (Fig. 1.2) coincide.

Figure 1.2 shows the possible interplay between the length scales. The small scales defined by the structure and the interactions need to be of the same order of magnitude in order to get a significant rate of adsorption and sticking, which will contribute to the reinforcement. The larger structures, such as agglomerates

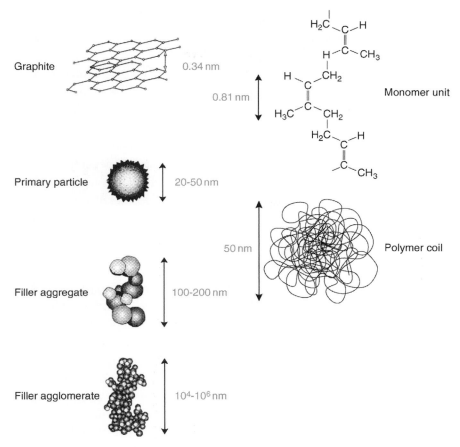

Fig. 1.2. Comparison of the different length scales for the elements. The filler particles, here carbon black, have basically carbon surfaces. These interact directly with the monomers on their length scales. However, the aggregates and agglomerates have dimensions similar to those of the polymer coils, so they can directly interact with them.

and aggregates have similar scales with typical polymer radii. Thus we can expect scale-dependent contributions to the modulus based on the interactions between rubber matrix chains and filler particles.

On yet larger scales hydrodynamic reinforcement comes into play. The basic idea goes back to Einstein and his work on the viscosity [5]. He derived an equation for the enhancement of the viscosity of solutions when spherical particles are added. This is the well-known formula

$$\eta = \eta_0(1 + 2.5\phi) , \tag{1.1}$$

where η_0 is the viscosity for the pure solution and ϕ is the volume fraction of the added spheres. The number 2.5 is purely geometrical and has its origin in the spherical nature of the added particles.

So far we have not mentioned the main contribution from the elastic matrix which comes in most cases from polymer networks, i. e. crosslinked polymer chains. The elasticity of such networks can be described on different levels (see the classical book of Treloar [6] for a basic reference). For the purpose of this book we restrict ourselves to the statistical physics description, i. e. simplified models are used which allow at least some of the molecular aspects to be taken into account. In physical terms the elastic modulus can be simply estimated: if a large number of chains become crosslinked by N_c crosslinks, each crosslink contributes with a thermal energy $k_B T$ to the elastic (free) energy. Thus the modulus in its simplest version should be of the form [7]

$$G_{\text{matrix}} \propto N_c k_B T \ . \tag{1.2}$$

As yet, the formation and structure of filler networks in elastomers and the mechanical response, e.g., the pronounced dynamic amplitude dependence or stress softening, of reinforced rubbers is not fully understood, though this question is of great technical interest. A deeper understanding of filler networking and reinforcement could provide a useful tool for the design, preparation and testing of high-performance elastomers, as applied in tires, seals, bearings, and other dynamically loaded elastomer components. In the past, attention has been primarily focussed on understanding the reinforcing mechanism of carbon black, the most widely used filler in the rubber industry [3, 8]. The strongly non-linear dynamic-mechanical response of carbon-black-filled rubbers, reflected primarily by the amplitude dependence of the viscoelastic complex modulus, was brought into clear focus by the extensive work of Payne [9–16]. Therefore, this effect is often referred to as the Payne effect.

As shown in Fig. 1.3 for a specific frequency and temperature, the storage modulus G' decreases from a small strain plateau value G'_0 to an apparently high amplitude plateau value G'_∞ with increasing strain amplitude. The loss modulus G'' shows a fairly pronounced peak. It can be evaluated from the tangent of the measured loss angle, $\tan \delta = G''/G'$, as depicted in Fig. 1.4. Obviously, the loss tangent shows a low plateau value at small strain amplitude, almost independent of filler concentration, and passes through a broad maximum with increasing strain.

Therefore we can expect that many different factors contribute to the modulus of a composite material. The contributions to the modulus from the different length and time scales are summarized schematically in Fig. 1.5.

The Payne effect of carbon black reinforced rubbers has also been investigated intensively by a number of different researchers [17–20]. In most cases, standard

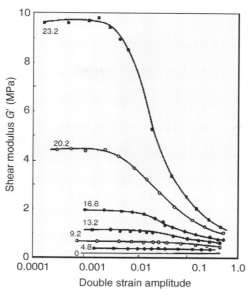

Fig. 1.3. Amplitude dependence of the storage modulus of butyl/N330 samples at various carbon black concentrations [9].

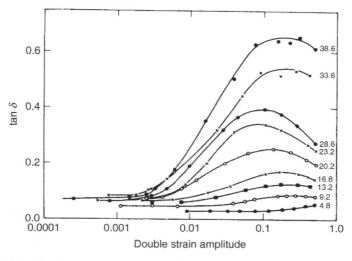

Fig. 1.4. Amplitude dependence of the loss tangent of the butyl/N330 samples shown in Fig. 1.3 at various carbon black concentrations [9].

diene rubbers that are widely used in the tire industry, such as styrene butadiene rubber (SBR), natural rubber (NR), and butadiene rubber (BR), have been employed, but carbon-black-filled bromobutyl rubbers [21–23] or functional rubbers containing tin end-modified polymers [24] have also been used. The Payne effect

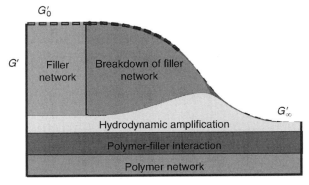

G'_0

G'

Filler network

Breakdown of filler network

G'_∞

Hydrodynamic amplification

Polymer-filler interaction

Polymer network

Log (strain amplitude γ_0)

Fig. 1.5. Different contributions on different length scales build up the modulus of the material.

was described in the framework of various experimental procedures, including preconditioning-, recovery- and dynamic stress-softening studies [25]. The typically almost reversible non-linear response found for carbon black composites has also been observed for silica-filled rubbers [25–27].

The temperature dependence of the Payne effect has been studied by Payne and other researchers [9, 13, 28]. With increasing temperature an Arrhenius-like drop of the moduli is found if the deformation amplitude is kept constant. As well as this effect, the impact of filler surface characteristics on the non-linear dynamic properties of filler reinforced rubbers has been discussed in a review of Wang [28], where basic theoretical interpretations and modeling are presented. The Payne effect has also been investigated in composites containing polymeric model fillers, like microgels of different particle size and surface chemistry, which could provide more insight into the fundamental mechanisms of rubber reinforcement by colloidal fillers [29, 30].

The pronounced amplitude dependence of the complex modulus, referred to as the Payne effect, has also been observed in low-viscosity media, e.g., composites of carbon black with decane and liquid paraffin [31], carbon black suspensions in ethylene vinylacetate copolymers [32], and for clay–water suspensions [33, 34]. It was found that the storage modulus decreases with dynamic strain amplitude in a qualitative manner similar to that for carbon-black-filled rubbers. This emphasizes the role in the Payne effect of a physically bonded filler network structure, which governs the small strain dynamic properties even in absence of rubber. Further, these results indicate that the Payne effect is primarily determined by structure effects of the filler. The elastomer seems to act merely as a dispersing medium that influences the kinetics of filler aggregation, but does not have a pronounced influence on the overall mechanical behavior of three-dimensional filler networks. However, the critical strain amplitude at which the Payne effect appears is found to be shifted to significantly smaller values if low-viscosity composites are used in place of rubber composites: This indicates a strong impact of the polymer matrix on the stability and strength of filler networks.

The strong non-linearity of the viscoelastic modulus with increasing dynamic strain amplitude has been related to a cyclic breakdown and reaggregation of filler–filler bonds [29, 35–37]. Thereby, different geometrical arrangements of particles in a particular filler network structure, resulting, e.g., from percolation as in the model of Lin and Lee [37] or kinetic cluster–cluster aggregation [29], have been considered. Nevertheless, a full micromechanical description of energy storage and dissipation in dynamically excited reinforced rubbers is still lacking.

As well as the Payne effect, which is relevant for dynamical loading of filler reinforced rubbers, the pronounced stress softening, which is characteristic of quasi-static deformations up to large strain, is of major interest for technical applications. This stress softening is often referred to as Mullins effect due to the extensive studies of Mullins and coworkers [38–40] on the phenomenon. Depending on the history of straining, e.g., the extent of previous stretching, the rubber material undergoes an almost permanent change that alters its elastic properties and increases hysteresis drastically. Most of the softening occurs in the first deformation and after a few deformation cycles the rubber approaches a steady state with a constant stress–strain behavior. The softening is usually only present at deformations that are smaller than the previous maximum. An example of (discontinuous) stress softening is shown in Fig. 1.6, where the maximum strain is increased, successively, from one uniaxial stretching cycle to the next.

The micromechanical origin of the Mullins effect is not yet fully understood [3, 17, 41]. In addition to the action of the entropy elastic polymer network, which is quite well understood on a molecular-statistical basis [42, 43], the impact of filler particles on stress–strain properties is of great importance. On the one hand the addition of hard filler particles leads to a stiffening of the rubber matrix that can be described by a hydrodynamic strain amplification factor [44–46]. On the other hand the constraints introduced into the system by filler–polymer bonds result in a decreased network entropy. Accordingly, the free energy, which equals the negative entropy times the temperature, increases linearly with the effective number of network junctions [44, 45, 47, 48]. A further effect is obtained from the formation of filler clusters or a filler network due to strong attractive filler–filler bonds [3, 17, 41, 44, 45, 47, 48].

Stress softening is supposed to be affected by different influences and mechanisms that have been discussed by a variety of authors. In particular, it has been attributed to a breakdown or slippage [49–52] and disentanglements [53] of bonds

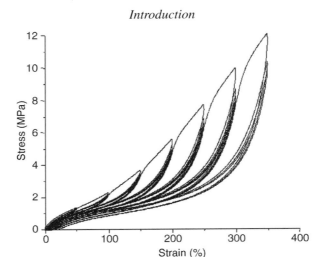

Fig. 1.6. Example of stress softening with successively increasing maximum strain after every fifth cycle for a solution SBR (S-SBR) sample filled with 50 phr carbon black.

between filler and rubber, a strain-induced crystallization–decrystallization [54, 55] or a rearrangement of network chain junctions in filled systems [40]. A model of stress-induced rupture or separation of network chains from the filler surface has been derived by Govindjee and Simo [50], who developed a complete macroscopic constitutive theory on the basis of statistical mechanics. A remarkable approach has been proposed by Witten *et al.* [56], who found a scaling law for the stress–strain behavior in the first stretching cycle by modeling the breakdown of a cluster–cluster aggregation (CCA) network of filler particles. They used purely geometrical arguments by referring to the available space for the filler clusters in strained samples, leading to universal scaling exponents that involve the characteristic fractal exponents of CCA clusters. However, they did not consider, though these are evident from experimental data, effects coming from the rubber matrix or the polymer–filler interaction strength e.g., the impact of matrix crosslinking or filler surface treatment (graphitization) on stress–strain curves. The stress softening indicates that stress-induced breakdown of filler clusters takes place, where the stress on the filler clusters is transmitted by the rubber matrix.

The above interpretations of the Mullins effect of stress softening ignore the important results of Haarwood *et al.* [54, 55], who showed that a plot of stress in the second extension versus the ratio between strain and prestrain of natural rubber filled with a variety of carbon blacks yields a single master curve [40, 54]. This demonstrates that stress softening is related to hydrodynamic strain amplification due to the presence of the filler. Based on this observation a micromechanical model of stress softening has been developed invoking hydrodynamic reinforcement of the rubber matrix by rigid filler clusters that are irreversibly broken during the first deformation cycle [57, 58]. Thereby, the extended tube model of rubber elasticity, introduced in Section 5.4, has been applied [42, 43, 59, 60]. This "dynamic flocculation model" is considered in Section 10.3.

The different contributions to the elastic modulus arise from completely different physical sources. However, it is not always clear how to separate the different contributions. Roughly, we can speak of the basic contributions to the modulus of a nano-composite system. The basis for the material is the elastic matrix, which in most cases is a highly elastic polymer network. Nevertheless, the symbolic diagram shown in Fig. 1.5 will serve as a model and a guideline throughout this book.

2

Basics about polymers

2.1 Gaussian chains – heuristic introduction

This chapter introduces a convenient view of the basic physics used in the description of polymer chains that will form a network which is the elastomer matrix.

In statistical polymer theory polymer chains are very simple objects. Of course, their local chemical structure can be very rich and many properties depend on the types of monomers which are used. Nevertheless, as the chain becomes longer, the specific monomers play a smaller and smaller role. The shape of the chain depends only on the environment rather than on any of the chemical details of the monomers. Therefore the simplest model to use for the present problem is that of a random walk. Although this model is very oversimplified, most polymers can be modeled in such a way [61]. The random walk model is very instructive here. First, it serves as a simple but instructive model for general problems in the statistical physics of polymers; second, it provides the basis for the simplest model of the elasticity of networks. We will turn to the latter point shortly.

Let us study the case of random walks in more detail and on a more formal basis [62]. To be more precise we start from the set of bond vectors $\{\mathbf{b}_i\}_{i=1}^N$, which are statistically independent vectors. The probability of finding a whole set is given by

$$P\left(\{\mathbf{b}_i\}_{i=1}^N\right) = \prod_{i=1}^N p(\mathbf{b}_i), \qquad (2.1)$$

where the probability $p(\mathbf{b}_i)$ is given by

$$p(\mathbf{b}_i) = \frac{1}{4\pi b^2}\delta(|\mathbf{b}_i| - b). \qquad (2.2)$$

The prefactor in (2.2) comes from the assumption of isotropy and corresponds to the normalization. We are now interested in finding the distribution of the end-to-end distance in order to make some statements about the size of the random walk. To do so, we remember the definition of the end-to-end distance $\mathbf{R} = \sum_{i=1}^{N} \mathbf{b}_i$ and compute its distribution:

$$P(\mathbf{R}) = \int \prod_{i=1}^{N} d\mathbf{b}_i \, \delta \left(\mathbf{R} - \sum_{i=1}^{N} \mathbf{b}_i \right) P \left(\{\mathbf{b}_i\}_{i=1}^{N} \right). \tag{2.3}$$

The calculation becomes simple if the delta function is parameterized by

$$\delta \left(\mathbf{R} - \sum_{i=1}^{N} \mathbf{b}_i \right) = \frac{1}{(2\pi)^3} \int d^3k \, \exp \left(-i\mathbf{k} \left(\mathbf{R} - \sum_{i=1}^{N} \mathbf{b}_i \right) \right). \tag{2.4}$$

Inserting this into (2.3) yields the classical Gaussian distribution function for a large number of monomers N:

$$P(\mathbf{R}) = \left(\frac{3}{2\pi b^2 N} \right)^{3/2} \exp \left(-\frac{3\mathbf{R}^2}{2b^2 N} \right). \tag{2.5}$$

Another important point to note is that the restriction of fixed length for the bond vectors can be relaxed without problems. Indeed, when the probability (2.2) is replaced by an effective Gaussian of the form

$$\tilde{p}(\mathbf{b}_i) = \left(\frac{1}{4\pi b^2} \right)^{3/2} \exp \left(-\frac{\mathbf{b}_i^2}{2b^2} \right), \tag{2.6}$$

there is no change in the result. In contrast to (2.2), where the bond length is constrained to take fixed values, (2.6) fixes only the mean squared distance between the two neighboring bonds.

Again we must note that (2.5) has a certain scaling function [63]. It contains two important pieces of information. To see this, let us rewrite it in the more convenient form

$$P(\mathbf{R}) = \frac{a}{\xi^3} \mathcal{F} \left(\frac{R}{\xi} \right). \tag{2.7}$$

Here we have introduced the only relevant scale in the problem $\xi = b\sqrt{N} \equiv bN^\nu$ (a is just a numerical constant). This scale, which corresponds of course to the size of the ideal polymer, spans a volume of ξ^3 in three space dimensions. There are two important observations. The first is that the distribution function depends only on the ratio R/ξ. The second is that the prefactor have the dimension of the volume, because of the normalization requirement, i.e. $\int d^3R \, P(\mathbf{R}) = 1$. Thus we might expect that we can make use of these facts for other polymeric objects, even though we cannot compute the analog of $P(\mathbf{R})$ completely. This is true if we take into account interactions and the chains become self-avoiding.

Before we proceed in these directions, we have to analyze (2.5) in more detail. A trivial observation is, as mentioned already, that the distribution function is purely Gaussian. This reflects once more that we had not taken any interactions into account. A second important point is that the distribution function $P(\mathbf{R})$ is invariant under any rescaling of the chain length, i. e. if N is replaced by $\tilde{N} = N/\lambda$, when λ is a real number. Of course, the numerical value of λ must be smaller than N itself, so that the rescaled chain can still be modeled as a random walk. Thus we must require $1 \leq \lambda << N$. A third point concerns the mean size of the polymer. The mean end-to-end distance is calculated as

$$\langle \mathbf{R}^2 \rangle = \int \mathrm{d}^3 R \, \mathbf{R}^2 P(\mathbf{R}) = b^2 N. \tag{2.8}$$

Of course, this model suffers from severe simplifications. For example, the chain does not interact with the environment. Even for a single chain model this is a dramatic simplification, since the chain can interact with itself. Therefore we have to find a way to include interactions in the model.

2.2 Gaussian chains – path integrals

The Gaussian chain is a very pedagogical example for the introduction of the path integral description of polymers. A Gaussian chain corresponds to a Feynman–Wiener path integral. Let us therefore present a heuristic argument [62]. Readers that are more interested in the mathematics should refer to the classical reference of Feynman and Hibbs [64], one of the best introductions to path integrals.

We have already noted that Gaussian chains are self-similar [62, 65, 66]. This point corresponds to the central limit theorem. To understand this we return to the Gaussian distribution for the mean size of the bond lengths.

$$p(\mathbf{b}_i) \cong \left(\frac{3}{2\pi b^2} \right)^{3/2} \exp \left\{ -\frac{3}{2b^2} \mathbf{b}_i^2 \right\}. \tag{2.9}$$

Of course, the distribution of the set is given by

$$P(\{\mathbf{b}_i\}) = \prod_{i=1}^{N} \left(\frac{3}{2\pi b^2} \right)^{3/2} \exp \left\{ -\frac{3}{2b^2} \mathbf{b}_i^2 \right\} \tag{2.10}$$

$$= \left(\frac{3}{2\pi b^2} \right)^{N3/2} \exp \left\{ -\frac{3}{2b^2} \sum_{i=1}^{N} \mathbf{b}_i^2 \right\}. \tag{2.11}$$

Now we recall that each bond vector is given by the difference of the spatial vectors of each bond, i. e. $\mathbf{b}_i = \mathbf{R}_i - \mathbf{R}_{i-1}$, and write the total probability as

$$P(\{\mathbf{b}_i\}) = \left(\frac{3}{2\pi b^2}\right)^{3N/2} \exp\left\{-\frac{3}{2b^2} \sum_{i=1}^{N} (\mathbf{R}_i - \mathbf{R}_{i-1})^2\right\}. \tag{2.12}$$

Formally we can associate a "Hamiltonian" with this expression. Indeed, if for a moment we write the distribution as

$$P(\{\mathbf{b}_i\}) = \mathcal{N} \exp\{-\beta H_0(\{\mathbf{R}_i\})\}, \tag{2.13}$$

we can define

$$\beta H_0 = \frac{3}{2b^2} \sum_{i=1}^{N} (\mathbf{R}_i - \mathbf{R}_{i-1})^2. \tag{2.14}$$

Throughout this section we will use $\mathcal{N}$ to denote any normalization factor that we do not want to determine precisely. From the above, we may recognize a well-known Hamiltonian; this Hamiltonian is used in solid state physics, to describe lattice vibrations of a one-dimensional solid as a chain of harmonic springs [67]. Crudely we may use the continuum limit

$$\frac{\mathbf{R}_i - \mathbf{R}_{i+1}}{1} \rightarrow \left(\frac{\partial \mathbf{R}}{\partial s}\right) \tag{2.15}$$

to arrive at a symbolic notation for the distribution

$$P(\{\mathbf{b}_i\}) = \mathcal{N} \exp\left\{-\frac{3}{2b^2} \int_0^N \left(\frac{\partial \mathbf{R}}{\partial s}\right)^2 ds\right\}. \tag{2.16}$$

Equation (2.16) is called the Wiener distribution (or in polymer theory the Wiener–Edwards distribution) for random walk chains. So far (2.16) does not contain anything new, except that a more fancy and more useful notation has been introduced. To go from discrete to continuous notation we can use the following minimal dictionary:

$$0 \le s \le N \Longleftrightarrow 1 \le i \le N \tag{2.17}$$

$$\int_0^N ds \Longleftrightarrow \sum_{i=1}^{N} i = 1. \tag{2.18}$$

Now we have to ask ourselves: what have we gained by reformulating the problem into this language? The advantage will become obvious. Equation (2.16) is written in a "language" that allows modern theoretical treatment by using functional integrals, which are well known in theoretical physics, especially in quantum mechanics.

Formally we can write for the partition function the symbolic expression

$$Z = \mathcal{N} \sum_{\text{all paths } \mathbf{R}(s)} \exp \left\{ -\frac{3}{2b^2} \int_0^N ds \left(\frac{\partial \mathbf{R}(s)}{\partial s} \right)^2 \right\}. \qquad (2.19)$$

The partition function is now represented as the sum over all possible paths. The physical idea behind this is as follows: All possible conformations of the random walk, which is composed of N statistical independent segments contribute to the value of Z. There are, of course, more probable paths and also less probable paths. An example of a less probable path is a stretched path. The appearance of an almost straight line with $R \propto N$ is very unlikely from entropic reasons, but nevertheless it contributes to the partition function Z. This mathematical formulation resembles the idea of path integrals in quantum mechanics. Indeed, we are going to build up a simple analogy to the Feynman representation of quantum mechanics [64].

To construct the analogy of "the sum over paths" we must realize first that the random walk polymer satisfies a diffusion equation. This becomes most obvious if we recall the distribution of the end-to-end distance $P(\mathbf{R})$ satisfies the diffusion equation [62, 64]

$$P(\mathbf{R}, N) = \left(\frac{3}{2\pi N b^2} \right)^{3/2} \exp \left\{ -\frac{3}{2Nb^2} \mathbf{R}^2 \right\}, \qquad (2.20)$$

which we have derived already. We interpret the equation in the following way. We want to construct all random walks between the space points $\mathbf{r}' = \mathbf{0}$, where the walk starts and the end point $\mathbf{r} = \mathbf{R}$. Additionally we require that the walker has N steps. It is easy to show that $P(\mathbf{R})$ satisfies a diffusion equation of the form

$$\left(\frac{\partial}{\partial N} - \frac{b^2}{6} \nabla^2 \right) P(\mathbf{R}, N) = 0 \qquad \forall \mathbf{R} \neq \mathbf{0}, N > 0. \qquad (2.21)$$

We can reformulate this in terms of a Green function for any two points $\mathbf{r}, \mathbf{r}'$ and corresponding contour variables s, s', i.e.

$$\left(\frac{\partial}{\partial s} - \frac{b^2}{6} \nabla^2 \right) G(\mathbf{R}, \mathbf{R}', s, s') = \delta(\mathbf{R} - \mathbf{R}') \delta(s - s'). \qquad (2.22)$$

The delta functions on the right-hand side of (2.22) are the initial conditions and ensure that the diffusion equation has only physical solutions for $s - s' \geq 0$, and that for $s - s' = 0$ we have $G(\mathbf{r}, \mathbf{r}', 0, 0) = \delta(\mathbf{r} - \mathbf{r}')$. The initial conditions therefore are

$$G\left(\mathbf{R}, \mathbf{R}', s, s'\right) = 0 \qquad \forall\left(s - s'\right) < 0 \,, \tag{2.23}$$

$$G\left(\mathbf{R}, \mathbf{R}', 0, 0\right) = \delta\left(\mathbf{R} - \mathbf{R}'\right) \,. \tag{2.24}$$

The solution of the differential equation is represented by the Green function of the random walk polymer

$$G\left(\mathbf{R}, \mathbf{R}', s, s'\right) = \left(\frac{3}{2\pi b^2 \left(s - s'\right)}\right)^{3/2} \exp\left\{-\frac{3}{2b^2} \frac{\left(\mathbf{R} - \mathbf{R}'\right)^2}{\left(s - s'\right)}\right\}. \tag{2.25}$$

2.3 Self-interacting chains

Gaussian chains are unrealistic in the sense that the segments may cross each other. Real polymer chains cannot do this and when two segments meet at the same place they have to repel each other. Thus, the most serious drawback of the models is that two chain segments are allowed to have the same coordinates $\mathbf{R}(s)$. In more realistic chain models this cannot happen. We must, however, introduce a repulsive potential $V\left(\mathbf{R}\left(s\right) - \mathbf{R}\left(s'\right)\right)$ [62, 66] which prevents the two monomers (or chain segments) being in the same place. To set up a better model we use a most plausible Hamiltonian for the self-avoiding walk chain. It is given by

$$\beta H\left(\{\mathbf{R}(s)\}\right) = \frac{d}{2b^2} \int_0^N \mathrm{d}s \left(\frac{\partial \mathbf{R}}{\partial s}\right)^2 + \frac{1}{2} \int_0^N \mathrm{d}s \int_0^N \mathrm{d}s' V\left(\mathbf{R}\left(s\right) - \mathbf{R}\left(s'\right)\right), \tag{2.26}$$

where we have now introduced the space dimension d as another parameter. We will see below, that this appears to be useful in some cases when we discuss the interactions in more general terms. The potential $V(\mathbf{r})$ is determined by the usual intramolecular potentials, such as the Lennard–Jones potentials, hard core interactions, etc., which are well known from the theory of liquids [68], but we will later use more simplified pseudopotentials. It has been shown that a useful pseudopotential approximation is [62, 63, 66]

$$V\left(\mathbf{R}\right) = v\delta\left(\mathbf{R}\right) \propto b^3 \delta\left(\mathbf{R}\right). \tag{2.27}$$

This potential is always repulsive as long as the chain segments are at the same place. The strength of the potential is roughly given by the excluded volume between two segments. This is of the order of b^3. We will see later that the precise value of v is not of significance with respect to the universal properties.

The first difficulty comes from the potential itself. In contrast to the considerations above, the excluded volume potential appears as a pair interaction. Therefore we cannot formulate it in terms of a simple diffusion equation. The first serious problem is therefore buried in the nature of the excluded volume: βH of a self-avoiding walk (SAW) does not correspond to a one-particle potential $\delta\left(\mathbf{R}\left(s\right)-\mathbf{R}\left(s'\right)\right)$.

The next serious problem appears if we try a perturbation theory that requires an expansion in terms of the excluded volume parameter v immediately rings alarm bells, i. e. if we work with an expansion of the form

$$G\left(\mathbf{R},\mathbf{R}',N\right)=G_0\left(\mathbf{R},\mathbf{R}',N\right)+v\left(\cdots\right)\pm v^2\left(\cdots\right),\qquad(2.28)$$

where $(\cdots)$ stands for expressions to be computed. We immediately see for this that the perturbation series diverges, which Fixman [69] was the first to realize that the perturbation parameter is not a small quantity. The perturbation parameter of relevance is not v itself, but the combination $v\sqrt{N},(v\sqrt{N})^2,(v\sqrt{N})^3,\cdots$ [62,66]. More generally in d dimensions the perturbation parameter is $vN^{(4-d)/2}$. The result on the chain size is (see e.g. [62])

$$\left\langle\mathbf{R}^2\right\rangle=Nb^2\left[1+\frac{4}{3}\left(vb^2N\right)^{(4-d)/2}+\text{ const. }\left(vb^2N\right)^{4-d}+\cdots\right].\qquad(2.29)$$

Thus any perturbation theory in $d<4$ must break down [66]. This means mainly that "new physics" beyond the random walk ideas takes over, and we cannot stay within the methods used so far. What will happen can be seen in a simple dimensional estimate of the Hamiltonian [70]. To resolve the problem of the diverging expansion terms for N to ∞ a dimensional argument can be proposed:

$$\beta H=\frac{d}{2b^2}\int_0^N ds\left(\frac{\partial\mathbf{R}}{\partial s}\right)^2+\frac{v}{2}\int_0^N ds\int_0^N ds'\delta\left(\mathbf{R}\left(s\right)-\mathbf{R}\left(s'\right)\right).\qquad(2.30)$$

The steps in the analysis are the following:

- suppose that the size of the polymer has scaling of the form $R\sim N^v$;
- estimate the connectivity term as $\sim N^{2v-2+1}$;
- estimate the excluded volume $\sim N^{2-dv}$;
- match both terms in the exponents: $2v-1=2-dv$ and read off the result

$$v=\frac{3}{2+d}.\qquad(2.31)$$

Here we see that the space dimension enters. Unlike for the random walk we can expect a dependence on the space dimension for the size if the chain is regarded

Table 2.1. *Flory type estimates for the critical exponents* ν

d	ν_F	comment
1	1	exact
2	3/4	exact
3	3/5	wrong
4	1/2	exact

as an SAW. Now we have to consider the quality of the results. The dimensional counting is crude, so we would not expect the results to be of use, but, when the estimates and the real values are compared, the result is a surprise. Let us first summarize the results in Table 2.1. The only dimension where the model goes wrong is $d = 3$. Let us discuss the results in the different dimensions in more detail. First, for $d = 1$ the result is exact, since the SAW in one dimension must be a fully stretched chain. $d = 1$ is the lowest critical dimension since ν cannot become larger than 1. Otherwise the chain would be overstretched. We just mention without proof that the value for $d = 2$ is also exact [66]. This has been proven by conformal invariance [71]. For $d = 3$ the result is close to the real value of $\nu = 0.589\ldots$, which has been computed by renormalization group theory.

We realize also that $\nu = 1/2$ for $d = 4$. Why is this special? We should not be too surprised, when we see that the perturbation parameter was estimated as $\nu N^{(4-d)/2}$. In dimensions larger than four, this parameter becomes really small. To be more precise look at a special Ginzburg argument and let us estimate the energy using

$$U = \frac{1}{2}\nu \int_0^N ds \int_0^N ds' \delta\left(\mathbf{R}\left(s\right) - \mathbf{R}\left(s'\right)\right) \propto \frac{N^2}{R^d}. \qquad (2.32)$$

If we put the ideal walk chain size in this equation, we get

$$U \sim \nu \frac{N^2}{R^d} \underbrace{\approx}_{R \propto \sqrt{N}} \nu N^{(4-d)/2}. \qquad (2.33)$$

Thus the SAW interaction is no longer important for $d \geq 4$ and we recover random walk behavior. The case $d = 4$ requires some attention. The exponent $\nu = 1/2$ is exact, but there are, however, logarithmic corrections to the prefactors and scaling functions. This can be seen intuitively, since the scaling estimate of the interaction potential is $U \propto N^0$, which in most cases indicates the existence of logarithmic corrections. These have been worked out in detail [66].

At the present level we are not able to compute the exponents more accurately. This requires more work, which we will outline in the next section. We can, however, use the scaling forms to find the asymptotic form of, for example, the distribution function. In the case of the random walk we found that the probability distribution was of a scaling form, see (2.7). We might assume that the SAW is also a self-similar object and that we can use the same argumentation. In doing so we might immediately guess the form [63]

$$P_{\text{SAW}}(\mathbf{R}, N) \propto \left(\frac{1}{N}\right)^{\nu d} \left(\frac{R}{bN^\nu}\right)^{(\gamma-1)/\nu} \exp\left\{-\left(\frac{R}{bN^\nu}\right)^{1/(1-\nu)}\right\}, \quad (2.34)$$

where γ is another exponent, which is $\gamma = 1$ for the random walk. For many more lucid discussions on these issues see the brilliant book by des Cloizeaux and Jannink [66].

3

Many-chain systems: melts and screening

3.1 Some general remarks

So far we have studied an isolated single chain in a good solvent, which corresponds to the case of the SAW. The most important result was the case of the swollen chain with the scaling law $R \sim N^{3/5}$. This introduces by inverting, in principle, a new fractal dimension $d_f = 5/3$ for the chain $R^{d_f} \sim N$. In the following we are going to study the problem of polymer melts or, correspondingly, concentrated polymer solutions. In other words we want to study the physical behavior of many-chain systems. What can we expect? To see this pictorially let us imagine a snapshot of a three-dimensional concentrated polymer solution (Fig. 3.1). Excluded volume correlations are now not only taking place within each single chain, but the increasing number of contact points with other chains at increasing polymer concentration result in additional excluded volume. At the same time the correlations within each chain are destroyed more and more. To some extent fewer correlations rule the statistical behavior of individual chains in the concentrated solution or the polymer melt. We will show below that these additional contacts have severe effects on the statistical behavior of the individual chains. The cartoon in Fig. 3.1 suggests the following behavior for highly concentrated systems. We must distinguish between (at least) two different length scales. One regime is given by $r \le \xi$. At these scales a chain piece experiences correlations only from itself, i. e. we may expect the classical self-avoiding behavior. For the other regime, $r \ge \xi$, the self-avoiding correlations do not play a significant role and we can expect chain statistics close to a Gaussian chain. From this naive picture we must conclude that ξ must be a function of the concentration. At this intuitive level we can already deduce one significant concentration, C^*, which characterizes the overlap between the chains. If the polymers just overlap, a chain occupies its own volume. Thus we have [63]

$$C^* = \frac{N}{R^d} = \frac{N}{N^{dv}} = N^{1-vd}. \tag{3.1}$$

This is an important result and we have to note that for large chain lengths N the overlap concentration C^* is very small.

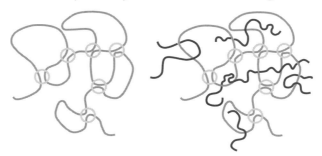

Fig. 3.1. The single chain versus a labeled chain in a melt. The single chain experiences contacts only with itself, whereas the melt chain has contacts with all neighboring chains. Thus the self-contacts, depicted by the circles, become irrelevant as the concentration increases. From [188], reprinted with permission of Elsevier.

3.2 Collective variables

We would like to formulate the problem more in terms of the chain model and the Edwards Hamiltonian. To begin, let us generalize the Edwards formulation to many chains. This is very simple and all that has to be done is to take into account the interactions between all the chain segments. This is reflected in the Hamiltonian [62, 72, 73]

$$
\beta H = \frac{3}{2b^2} \sum_{\alpha=1}^{n_p} \int_0^{N_\alpha} \left(\frac{\partial \mathbf{R}_\alpha}{\partial s} \right)^2 ds + \frac{1}{2} v \sum_{\alpha,\beta=1}^{n_p} \int_0^{N_\alpha} ds \int_0^{N_\beta} ds' \delta \left(\mathbf{R}_\alpha(s) - \mathbf{R}_\beta(s') \right),
$$

(3.2)

n_p the number of polymer chains present and all the other symbols have the same meaning as before. The principal task is to compute the partition function

$$
Z = \int \prod_{\alpha=1}^{n_p} \mathcal{D} \mathbf{R}_\alpha(s) \exp(-\beta H([\mathbf{R}_\alpha(s)])).
$$

(3.3)

Of course, this is generally not simple and the partition function cannot be computed exactly. Therefore a number of simplifications are necessary. The first one is to assume monodispersity which means that all chains have the same length. Mathematically this corresponds to $N_\alpha = N_\beta, \forall \alpha, \beta$. The next problem is that the partition function contains too many degrees of freedom. The number of chains n_p involved can be very large, and every chain itself has internal degrees of freedom, since they are assumed to be totally flexible. For these reasons it is convenient to introduce collective variables, which in this case are the polymer segment densities defined as

$$
\rho(\mathbf{x}) = \frac{1}{V} \sum_{\alpha=1}^{n_p} \int_0^N ds \delta(\mathbf{x} - \mathbf{R}_\alpha(s)).
$$

(3.4)

In fact, $\rho(\mathbf{x})$ can be viewed as a microscopic density operator whose value defines the density at an arbitrary point $\mathbf{x}$. It is therefore desirable to transform the Edwards Hamiltonian, which is a function of the real chain variables, to an effective one that depends only on the collective density variables. Let us therefore try a transformation, which is written formally as

$$H(\{\mathbf{R}_\alpha(s)\}) \longrightarrow \underbrace{H(\{\rho(\mathbf{x})\})}_{\text{effective Hamiltonian}} , \tag{3.5}$$

The resulting Hamiltonian is called "effective" here since it does not contain all the initial information. It can be imagined that the transformation cannot carried out exactly.

In the following we will show more of the transformation, since it has become an important tool in polymer physics. The technical strategy is quite simple. The strategy corresponds to the simple mathematical change of variables. The only difference is that it has to be carried out functionally. The result that we will aim for corresponds to the so-called random phase approximation (RPA), which has been frequently used in solid state physics. In the following we will not present the computation in detail but we outline the important steps. Some of the details can be found in [62, 73].

1. *Transformation to* **k**-*space* The first step is to use a formulation in reciprocal space. The advantage of this is that it simplifies the notation. To start, let us transform the density variable into **k**-space. This is very simple, and the result can be immediately written down:

$$\rho(\mathbf{x}) = \sum_{\alpha=1}^{n_p} \int_0^N ds \frac{1}{V} \sum_{\mathbf{k}} \exp\left[-i\mathbf{k}\left(\mathbf{x} - \mathbf{R}_\alpha(s)\right)\right] = \sum_{\mathbf{k}} \exp\left[-i\mathbf{k}\mathbf{x}\right]\rho_{\mathbf{k}}. \tag{3.6}$$

For the latter step we have simply used the Fourier transform of the density, i. e.

$$\rho_{\mathbf{k}} \equiv \frac{1}{V} \sum_{\alpha=1}^{n_p} \int_0^N ds \exp\left[i\mathbf{k}\cdot\mathbf{R}_\alpha(s)\right]. \tag{3.7}$$

One technical problem is how to treat the sum over all wave vectors $\mathbf{k}$. The exact enumeration can be carried out on a lattice, but it is useful to handle the sum over the wave vectors in its continuum version:

$$\sum_{\mathbf{k}} = \frac{V}{(2\pi)^d} \int d^d k. \tag{3.8}$$

The sum over the **k**-vectors appears very complicated, but is much simpler, if we note that the density must be a real number. Thus we make use of

$$\rho(\mathbf{x}) \in \mathbb{R} \Rightarrow (\rho_{\mathbf{k}})^* = \rho_{-\mathbf{k}} \tag{3.9}$$

in the following and realize that only a certain number of values will contribute to the sum, i.e. only $\mathbf{k} > \mathbf{0}$ are independent. In (3.9) the "*" denotes the complex conjugate.

2. *Transformation of variables* The second step is the most technical one. Here we have to transform the Hamiltonian from the chain variables $\mathbf{R}_\alpha(s)$ to the collective variables $\rho_\mathbf{k}$. The computation is very involved and we are not going to write all details here, but instead concentrate on the main issues. Formally we may write the transformation as

$$H\left(\{\mathbf{R}(s)\}\right) \longrightarrow H\left(\{\rho_\mathbf{k}\}\right), \tag{3.10}$$

and it becomes clear that this cannot be carried out exactly. Moreover, we will see later that we can only go in the direction of the arrow in (3.10). Thus the transformation cannot be inverted. The first formal step is to use the identity for the partition function:

$$Z = \int \prod_\alpha \mathcal{D}\mathbf{R}_\alpha(s) \underbrace{\int \prod_\mathbf{k} \mathrm{d}\rho_\mathbf{k} \delta\left(\rho_\mathbf{k} - \hat{\rho}_\mathbf{k}\right)}_{\equiv 1} \exp\left[-\beta H(\mathbf{R}_\alpha(s))\right]. \tag{3.11}$$

Here we have just inserted 1 which is expressed as a complicated functional integral over density variables. The density operators $\hat{\rho}$ correspond to $\rho_\mathbf{k}$ in (3.7). This expression contains all monomer positions $\mathbf{R}_\alpha(s)$, which we want to remove. The common way to proceed is to use a functional Fourier representation of the delta function $\delta\left(\rho - \hat{\rho}\right)$ in the form

$$\int \prod_\mathbf{k} \mathrm{d}\rho_\mathbf{k} \delta\left(\rho_\mathbf{k} - \frac{1}{V}\sum_\alpha \int_0^N \mathrm{d}s\, \exp\left[\mathrm{i}\mathbf{k}\cdot\mathbf{R}_\alpha(s)\right]\right) \equiv$$

$$\equiv \int \prod_\mathbf{k} \mathrm{d}\rho_\mathbf{k} \int \prod_\mathbf{k} \mathrm{d}\phi_\mathbf{k} \exp\left\{-\mathrm{i}\sum_\mathbf{k} \phi_{-\mathbf{k}}\hat{\rho}_\mathbf{k}\left(\rho_\mathbf{k} \sum_{\alpha=1}^{n_p} \int_0^N \mathrm{d}s\, \exp\left[\mathrm{i}\mathbf{k}\cdot\mathbf{R}_\alpha(s)\right]\right)\right\}, \tag{3.12}$$

where we have introduced an auxiliary field $\phi_\mathbf{k}$ for each value of the (formally) discrete wave vector. This auxiliary field parameterizes each of the terms in the product of delta functions in (3.12).

3. *Putting together and exchanging integration* We are now in the position to compute the partition function. To do so, we put the parameterized delta functions into the partition function and interchange the order of the integration. Thus we write the partition function in the following order:

$$Z = \int \prod_\mathbf{k} \mathrm{d}\rho_\mathbf{k} \int \prod_\mathbf{k} \mathrm{d}\phi_\mathbf{k} \exp\left\{-\mathrm{i}\sum_\mathbf{k} \phi_{-\mathbf{k}}\rho_\mathbf{k}\right\}$$

$$\times \int \prod_{\alpha=1}^{n_p} \mathcal{D}\mathbf{R}_\alpha(s) \exp\left\{-\beta H\left(\{\mathbf{R}_\alpha(s)\}\right) + \mathrm{i}\sum_\mathbf{k} \phi_\mathbf{k}\hat{\rho}_\mathbf{k}\right\}. \tag{3.13}$$

The main advantage of (3.13) is that the term on the second line depends only on the auxiliary field variables ϕ_k after the integrations over the chain variables are carried out. These integrations cannot be carried out exactly as mentioned earlier. Mathematically the second term contains the Jacobian of the variable transformation and physically it corresponds to a Legendre transformation of the original partition function.

Another essential point is that the exponential in the first line can be written as

$$\sum_k \phi_{-k} \hat{\rho}_k = \frac{1}{V} \sum_k \phi_{-k} \sum_\alpha \int_0^N ds \exp\left[i\mathbf{k} \cdot \mathbf{R}_\alpha(s) \right]$$

$$\equiv \sum_\alpha \int_0^N ds \phi\left(\mathbf{R}_\alpha(s)\right). \tag{3.14}$$

Thus each chain can be formally seen in a (random) field $\phi(\mathbf{R}_\alpha(s))$. The excluded volume pair interaction has been transformed to a one-particle interaction, i.e. an external field.

Now we have to go in a different direction. Instead of setting up a proper field theory and using a Schrödinger-type equation we stay in the real-space formulation and perform the $\mathbf{R}_\alpha$ integration. This procedure yields an effective Hamiltonian of the symbolic form $H(\rho_k, \phi_k)$, i.e. it depends only on two variables: ρ_k and the auxiliary field ϕ_k. The next step is to integrate out the remaining ϕ_k auxiliary variables. Indeed the ϕ-integration produces in a symbolic notation a Hamiltonian that depends only on the ρ_k variables:

$$Z_{\text{eff}} = \int \prod_k d\rho_k \exp\left(-\beta H_{\text{eff}}(\rho_k)\right). \tag{3.15}$$

This defines the desired Hamiltonian via the partition function. Nevertheless the strategy has now made it clear that we wish to go one step further and study the structure of both "Hamiltonians" $H(\rho_k, \phi_k)$ and $H_{\text{eff}}(\rho_k)$ in more detail. Below we will show that the auxiliary variable ϕ_k has a physical meaning, although it has been introduced just to parameterize the delta function during the change of the variables from $\mathbf{R}_\alpha(s) \rightarrow \rho_k$.

To begin with let us apply the procedure to the problem we would like to study. The transformation of the many-chain Hamiltonian is the first step. The part of the interactions, i.e. the mutual self-avoidance between all chains, is very simple. We write down again the starting point for the many-chain problem:

$$\beta H\left(\{\mathbf{R}_\alpha(s)\}\right) = \frac{3}{2b^2} \sum_{\alpha=1}^{n_p} \int_0^N ds \left(\frac{\partial \mathbf{R}_\alpha(s)}{\partial s}\right)^2$$

$$+ \frac{v}{2} \underbrace{\sum_{\alpha\beta} \int_0^N ds \int_0^N ds' \delta\left(\mathbf{R}_\alpha(s) - \mathbf{R}_\beta(s')\right)}_{\sum_k \rho_k \rho_{-k}}. \tag{3.16}$$

We then see that the Hamiltonian always has the general structure

$$\beta H_{\text{eff}}(\rho_{\mathbf{k}}) = \beta H_0(\rho_{\mathbf{k}}) + \frac{v}{2}\sum_{\mathbf{k}} \rho_{\mathbf{k}}\rho_{-\mathbf{k}}\,, \tag{3.17}$$

as long as only two-body interactions are of importance. The next task is the determination of H_0.

4. *Determination of H_0* To do this we have to consider the following integral for A_0:

$$A_0 = \int \prod_{\alpha=1}^{n_{\text{p}}} \mathcal{D}\mathbf{R}_\alpha(s)\exp\left\{-\frac{3}{2b^2}\sum_\alpha \int_0^N \left(\frac{\partial\mathbf{R}_\alpha}{\partial s}\right)^2\right.$$
$$\left. + \mathrm{i}\sum_{\mathbf{k}}\phi_{-\mathbf{k}}\sum_\alpha \int_0^N \mathrm{d}s\,\exp\left[\mathrm{i}\mathbf{k}\cdot\mathbf{R}_\alpha(s)\right]\right\}. \tag{3.18}$$

Formally A_0 corresponds to the partition function of a set of n_{p} polymers in a random field $\phi(\mathbf{R}_\alpha(s))$. This is a well-posed problem with which we had already dealt. We now see the real advantage of this procedure: The problem is now diagonal in all monomer indices, i.e. there are no couplings between different monomers s, s' and α, β. Next we carry out the $\mathbf{R}_\alpha(s)$-integration. We will then be left with an expression which depends only on the auxiliary variable. Of course, this is, in general, only possible using approximations, the most important of which is the assumption of small fluctuations in variables $\rho_{\mathbf{k}}$ and $\phi_{\mathbf{k}}$. This turns out to be consistent with the assumption of dense systems. In fact, the larger the polymer density is, the smaller the fluctuations are, and hence, the better the assumptions of small fluctuations. Intuitively this can easily be imagined. In dense melts the density fluctuations are much less pronounced, compared with a dilute solution, just because of the space-filling fraction of the polymers. In low-concentration solutions the spatial fluctuations are given by the single-chain conformations, whereas in melts the scales of the individual chain sizes do not play a major role, and fluctuations in the density are less pronounced.

The above assumption allows cumulant expansion of the integral. To simplify notation, we use the operator form of the collective density, i.e. (3.18) takes the more convenient form

$$A_0 = \int \prod_{\alpha=1}^{n_{\text{p}}} \mathcal{D}\mathbf{R}_\alpha(s)\exp\left\{-\frac{3}{2b^2}\int_0^N\left(\frac{\partial\mathbf{R}}{\partial s}\right)^2 + \mathrm{i}\sum_{\mathbf{k}}\phi_{-\mathbf{k}}\hat\rho_{\mathbf{k}}\right\}. \tag{3.19}$$

To proceed we try to approximate (3.19) in the following form:

$$A_0 \simeq \exp\left\{-\frac{1}{2}\sum_{\mathbf{k}\mathbf{k}'}\phi_{-\mathbf{k}}\Gamma^{(2)}_{\mathbf{k}\mathbf{k}'}\phi_{\mathbf{k}'}\right.$$
$$\left. - \sum_{\mathbf{k}_1,\mathbf{k}_2,\mathbf{k}_2}\phi_{\mathbf{k}_1}\phi_{\mathbf{k}_2}\Gamma^{(4)}(\mathbf{k}_1,\mathbf{k}_2,\mathbf{k}_3)\phi_{\mathbf{k}_3}\phi_{-\mathbf{k}_1-\mathbf{k}_2-\mathbf{k}_3} \pm \text{etc}\right\}. \tag{3.20}$$

This can be done and it turns out that the lowest-order function Γ is given by

$$\Gamma_{\mathbf{kk'}}^{(2)} = \langle \hat{\rho}_{\mathbf{k}} \hat{\rho}_{\mathbf{k'}} \rangle_0 \tag{3.21}$$

The average here is defined as

$$\langle \cdots \rangle_0 = \mathcal{N} \int \prod_{\alpha=1}^{n_p} \mathcal{D}\mathbf{R}_\alpha \, (\cdots) \exp\left\{ -\frac{3}{2b^2} \sum_{\alpha=1}^{N} \int_0^N \left(\frac{\partial \mathbf{R}_\alpha}{\partial s}\right)^2 \right\}, \tag{3.22}$$

where $\mathcal{N}$ is an appropriate normalization. At this point the advantage of the expansion can be seen. All the contributions to βH_0 can be expressed as averages of the ideal chain conformation, i.e. the non-interacting random walk. Thus the terms can be evaluated very simply. Obviously, the lowest-order contribution corresponds to the structure factor of the ideal chain

$$\Gamma_{\mathbf{kk'}}^{(2)} \equiv \underbrace{S^0(\mathbf{k})}_{\text{bare structure factor}} \delta\left(\mathbf{k} - \mathbf{k'}\right), \tag{3.23}$$

where S^0 is the base structure factor. Let us stick to the Gaussian order, which becomes reasonable for concentrated polymer solutions and polymer melts, where the fluctuations are very small. In this limit everything can be calculated on a simple level. The effective Hamiltonian $H[\rho_{\mathbf{k}}, \phi_{\mathbf{k}}]$ becomes to Gaussian order

$$H[\rho_{\mathbf{k}}, \phi_{\mathbf{k}}] = v \sum_{\mathbf{k}} \left\{ \frac{v}{2} |\rho_{\mathbf{k}}|^2 + \frac{1}{2} S^0(\mathbf{k}) |\phi_{\mathbf{k}}|^2 \right\}$$
$$+ \sum_{\mathbf{k}} i\phi_{\mathbf{k}} \rho_{-\mathbf{k}}$$
$$+ \text{ higher order in } \phi_{\mathbf{k}}. \tag{3.24}$$

Equation (3.24) appears not to be a real Hamiltonian, because at first sight it contains complex contributions. This is a somewhat confusing notation, but the problem can be resolved immediately, if the conjugate part is added. As before, the term containing the imaginary unit i must be replaced by

$$\phi_{-\mathbf{k}} \rho_{\mathbf{k}} \rightarrow \frac{1}{2} \left(\phi_{-\mathbf{k}} \rho_{\mathbf{k}} + \phi_{\mathbf{k}} \rho_{-\mathbf{k}} \right),$$

and then will be well defined and the exponent will be real. Only then are all the averages well defined.

5. *Gaussian model* Now we do the final ϕ-integration at the Gaussian level, i.e. all higher orders of the expansion in terms of the auxiliary field are neglected. Thus we start from the expression for the effective Hamiltonian

$$\beta H_{\text{eff}}[\rho_{\mathbf{k}}] = \frac{v}{2} \sum_{\mathbf{k}} \left(\frac{1}{S^0(\mathbf{k})} + v \right) \rho_{\mathbf{k}} \rho_{-\mathbf{k}} + \mathcal{O}\left(\rho^3, \rho^4, \ldots\right) \tag{3.25}$$

and compute the corresponding averages. The most important of these is the structure factor of the interacting system. The computation is trivial and starts from the definition of the structure factor as the density correlation function,

$$S(\mathbf{k}) = \langle \rho_{\mathbf{k}} \rho_{-\mathbf{k}} \rangle. \tag{3.26}$$

Again we recall the definition of the averages in terms of the collective variables, i. e.

$$\langle \cdots \rangle = \frac{1}{Z_{\text{eff}}} \int d\rho_{\mathbf{k}} \rho_{\mathbf{k}} \rho_{-\mathbf{k}} \exp(-H_{\text{eff}}), \tag{3.27}$$

where Z_{eff} is given by (3.15). Equation (3.27) yields the celebrated equation for the structure factor of a concentrated polymer solution

$$\frac{1}{S(\mathbf{k})} = \frac{1}{S^0(\mathbf{k})} + v. \tag{3.28}$$

Equation (3.28) is often called the standard RPA result for a polymer melt [62, 63]. It is remarkable, that the RPA result works very well, despite the crude approximations, and (3.28) has been confirmed by experiments with dense polymer solutions. In fact, we will come back later to this equation in the context of polymer blends and copolymers.

We could stop here because we have achieved what we wanted. The transformation of the Hamiltonian to collective variables is complete, and the structure factor is computed, at least in lowest order. However, we have still some unsolved questions, even at this level of the approximation. The next questions we want to look at are: does the auxiliary field have a physical meaning? And what does this theory so far mean for the conformation of chains in melts and concentrated solutions? We have already found that the single-chain correlations are destroyed as the polymer concentration increases. Can we then expect effects on the size of a labeled or tagged chain?

3.3 The statistics of tagged chains

So far we discussed the behavior of a dense polymer solution in terms of collective properties, such as the structure factor and the scattering properties of the system. One factor which we have not addressed yet is the behavior of the chains in the melt. In the introductory remarks we thought about the statistics of chains in the solution and the melt. We guessed that the size of the chain cannot be ruled by excluded volume forces alone, as in the case of isolated chains, since additional correlations from other chains also play an important role. This question is fortunately connected to another formal one.

We introduced an auxiliary field to represent the delta function when we changed the variables. This was a very formal point but a legitimate question is: does the auxiliary field $\phi_{\mathbf{k}}$ have a physical meaning? The answer is, of course, yes. To see this let us compute the correlator $\langle \phi_{\mathbf{k}} \phi_{-\mathbf{k}} \rangle$ in the lowest Gaussian order. Again the calculation is trivial, but very instructive, and the result is

$$\langle \phi_{\mathbf{k}} \phi_{-\mathbf{k}} \rangle = \frac{v}{1 + v S^0(\mathbf{k})} \equiv U . \tag{3.29}$$

This result was first found by Edwards [73] in a different form and by a different calculation. A simple dimensional analysis shows that U must be a renormalized interaction. This can be seen from its units: it must be the same as the "bare" excluded volume parameter v. It is instructive to bring (3.29) into the more appropriate form

$$\frac{1}{U} = \frac{1}{v} + S^0(\mathbf{k}). \tag{3.30}$$

This result yields that the original excluded volume interaction becomes renormalized in the presence of the other chains and the renormalized interaction in the melt is smaller than the bare interaction. However, the interaction becomes screened out. Edwards [73] used a different route. He derived the same result for the potential by using collective variables and integrating over all chains (in terms of collective variables), except one. Then he was left with an effective chain in the melt. The form he derived was

$$U(\mathbf{k}) = v - \frac{v^2}{\dfrac{1}{S^0(\mathbf{k})} + v}, \tag{3.31}$$

which agrees with (3.29) and (3.30). This form of the effective interaction is very instructive: The bare excluded volume v is reduced by a term stemming from the collective properties. We now need a physical picture and an interpretation of this point.

Therefore, we are going to have look at the effective $\rho_{\mathbf{k}}$-Hamiltonian, i.e. what is left when we have integrated out the $\phi_{\mathbf{k}}$-fields in (3.24). By some simple manipulation we see

$$
\begin{aligned}
H_{\text{eff}}[\rho_{\mathbf{k}}] &= \frac{v}{2} \sum_{\mathbf{k}} \left(\frac{1}{S^0(\mathbf{k})} + v \right) \rho_{\mathbf{k}} \rho_{-\mathbf{k}} \\
&= \frac{v}{2} \sum_{\mathbf{k}} \left(\underbrace{\frac{1}{Nc}}_{\text{ignore}} + \frac{k^2 b^2}{12c} + v \right) \rho_{\mathbf{k}} \rho_{-\mathbf{k}} \\
&\cong \frac{v}{2} \sum_{\mathbf{k}} \frac{b^2}{12c} \left(k^2 + \xi^{-2} \right) \rho_{\mathbf{k}} \rho_{-\mathbf{k}},
\end{aligned}
\tag{3.32}
$$

which are appropriate forms for the physical interpretation. The steps of the small computation involve, first, a Padé approximation for the structure factor and, second, neglect of the small $(1/N)$-term. In the last line we introduced a screening length

$$\xi = \left(\frac{b^2}{12cv}\right)^{1/2}. \tag{3.33}$$

Why did we call the length scale a screening length? To see this let us again compute the structure factor in terms of the length ξ. In this notation this is given by

$$S(\mathbf{k}) \simeq \frac{12c}{b^2} \frac{1}{k^2 + \xi^{-2}} \tag{3.34}$$

and can be immediately transformed back in three-dimensional real space. The structure factor in real space is then

$$S(\mathbf{r}) = \frac{3c}{\pi b^2} \frac{1}{r} \exp\left(-r/\xi\right), \tag{3.35}$$

which is of the Yukawa type. We see that the ξ plays the role of a characteristic screening length. At scales below ξ, the structure factor shows strong $1/r$ correlation. At larger scales, $r > \xi$, the correlation falls off exponentially, i. e. it is screened out. The screening length depends strongly on the concentration c, which becomes smaller with increasing concentration. This means that the strong correlations are destroyed faster at smaller length scales. In other words, we can say that the interactions become screened by increasing concentration.

To confirm this idea of screening interactions it is useful to calculate Hamiltonian $H(\mathbf{R}_T(s))$ for a test chain in the concentrated solution. To do so, we represent the n_p chain Hamiltonian by the collective variables described above. Into this medium we can insert an additional chain of the same chain length N which we call $\mathbf{R}_T(s)$. Of course, this test chain interacts with itself and the medium. The interaction term with the medium represents a coupling between the test chain and the medium. Thus we may represent the Hamiltonian for the test chain and the medium simply by

$$\beta H(\mathbf{R}_T(s)) = \frac{3}{2b^2} \int_0^N ds \left(\frac{\partial \mathbf{R}_T(s)}{\partial s}\right)^2 + \frac{v}{2} \int_0^N \int_0^N ds ds' \delta\left(\mathbf{R}_T(s) - \mathbf{R}_T(s')\right)$$
$$+ v \int_0^N ds \rho(\mathbf{R}_T(s))$$
$$+ \frac{v}{2} \sum_{\mathbf{k}} \left\{\frac{1}{S^{(0)}(\mathbf{k})} + v\right\} \rho_{\mathbf{k}} \rho_{-\mathbf{k}} + \mathcal{O}(\rho_{\mathbf{k}}^4). \tag{3.36}$$

Here the first line of the equation represents the bare Hamiltonian of the tagged chain, the second line is the coupling to the medium, i. e. the melt composed of the same chains, and the third term represents the polymer melt (medium) itself. Again, if higher orders in $\rho_{\mathbf{k}}$ are neglected the collective variables can be integrated out and we are left with an effective Hamiltonian for the test chain:

$$\beta H\left[\mathbf{R_T}(s)\right] = \frac{3}{2b^2} \int_0^N ds \left(\frac{\partial \mathbf{R}_T}{\partial s}\right)^2 + \frac{1}{2} \int_0^N ds \int_0^N ds' U\left(\mathbf{R}_T(s) - \mathbf{R}_T(s')\right),$$
(3.37)

where the potential $U(\mathbf{r})$ is called the "screened potential." It has the following explicit form in $\mathbf{k}$-space:

$$U(\mathbf{k}) = \frac{v\,k^2}{k^2 + \frac{1}{\xi^2}} = \frac{v}{1 + S^0(\mathbf{k})v} \equiv \langle \phi_k \phi_{-\mathbf{k}} \rangle.$$
(3.38)

The agreement with the correlation function of the $\phi_{\mathbf{k}}$ fields is obvious and shows that the previous guess is correct. It is instructive to calculate its Fourier transform to real space, which is

$$U(\mathbf{R}) = v \left[\underbrace{\delta(\mathbf{R})}_{\text{original ev}} - \underbrace{\frac{1}{4\pi\xi^2} \frac{e^{-R/\xi}}{R}}_{\text{screening}} \right].$$
(3.39)

Equation (3.39) shows explicitly why the melt potential is called the "screened potential." The bare excluded volume interaction v becomes screened out by the presence of the other chains. The range of interaction is mainly given by the screening length ξ. Moreover, at large values of r, or equivalently at $k = 0$, the interaction is zero. This leads to the conclusion that chains in dense polymer solution or in polymer melts will somehow have Gaussian statistics, i. e. their mean size R will obey Gaussian scaling, $R \propto \sqrt{N}$. This can be seen by considering a simple perturbation analysis for the size of the chain, where the perturbation is the effective potential $U(\mathbf{r})$. This calculation is straightforward and we only quote the result. The size of the chain is

$$\left\langle R^2 \right\rangle \cong Nb^2 \left[1 + \frac{12}{\pi} \frac{v}{b^4}\xi\right].$$
(3.40)

Equation (3.40) shows that the chain behaves like a Gaussian with respect to scaling with the chain length, but with a renormalized prefactor. The prefactor contains the screening length ξ and is thus concentration-dependent. It is also easy to show that all higher-order terms in the perturbation analysis are of less importance.

The next step is to develop a physical picture of this formal result. Obviously the screening can be understood by introducing a blob size which comes from the

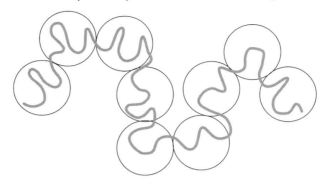

Fig. 3.2.   The blob picture of chains in concentrated polymer solutions. The chain can be divided into blobs of size ξ. Inside the blobs, the chain exhibits SAW behavior. Outside the blobs, the SAW correlations are destroyed and the chain becomes Gaussian. From [65], reprinted with permission.

screening length ξ. From (3.40) we see explicitly that the exponent in melts is $\nu = \frac{1}{2}$ and the chain is no longer self-avoiding. Higher-order perturbations are of course smaller and not N-dependent as in the SAW case. This allows the "blob picture" to be confirmed. Indeed the chain can be replaced by an effective chain of blobs. These blobs have a diameter of the order of ξ. Inside the blobs SAW correlations dominate the physical behavior, i. e. the chain is expected to be SAW-like. Outside the blobs, the many-chain correlations destroy the SAW character, and as we have just shown in (3.40) the chain becomes Gaussian (see Fig. 3.2). We have, however, made a mistake somewhere. This becomes obvious if we look at (3.35). We have just required that inside the blob there should be no volume correlations, therefore we should expect a structure factor $S(\mathbf{r}) \propto r^{-4/3}$ in three dimensions, since in reciprocal space this factor scales as $S(\mathbf{k}) \propto k^{-5/3}$ in $d = 3$. Equation (3.35), however, shows only a scaling with the inverse of the distance for scales $r < \xi$. Our mistake becomes immediately clear, when we remember that we had worked these results out only in the Gaussian approximation and we neglected all higher-order terms in $\phi_{\mathbf{k}}$ or $\rho_{\mathbf{k}}$. However, it is possible to recover the correct scaling by higher-order expansions together with renormalization theory [66]. The technical details of these computations are beyond the scope of the present chapter, but we will come back to this problem when we consider scaling theory. Shortly, we will put forward physical arguments that yield the desired results without doing too much technical work. This is the great advantage of scaling theory: simple physical pictures together with physical intuition quickly yield the first-order results.

However, dry rubber consists of many chains. Their interactions are of the same nature as those in polymer melts; their excluded volume interactions in the rubber matrix are screened.

4

Rubber formation

4.1 Classical theory of gelation

In this chapter we are going to summarize briefly the formation of the rubber matrix. The processes are strongly non-ergodic and often non-equilibrium. We will also see that some of the structural elements are even determined by the process of gelation and vulcanization; for the classical theories, see [74, 75] and for later developments see the book by Stauffer [76]. Any change from a melt to a rubber structure no matter by which process is a liquid-to-solid transition, which is monitored by a strong increase of the viscosity, as more material is connected together. If (formally) the viscosity tends to infinity at a certain conversion, a finite shear modulus can be measured. Indeed, this corresponds to a real phase change from a liquid phase to a solid phase under the change of the transport and mechanical properties. Of course, the critical point, at which the viscosity is infinite and the modulus is non-zero for the first time, depends on the number of crosslinks introduced into the melt. Therefore it is proposed that the viscosity increases by a power law:

$$\eta \propto (p_{\mathrm{c}} - p)^{-k},$$ (4.1)

and correspondingly the modulus increases by

$$G \propto (p - p_{\mathrm{c}})^{t},$$ (4.2)

where p_{c} is the "critical point" or in other words the critical conversion where the transition from the liquid- to solid-like behavior occurs [76].

The simplest model for such a so-called sol–gel transition was first proposed by Flory and Stockmayer. In this it was simply assumed that a large number of φ-functional molecules can be connected to a large (even macroscopically large) three-dimensional molecule by chemical reactions [74, 75]. At reaction time zero, the molecules form nothing but a melt. Then certain bonds in the side-chains of

31

the molecules react and clusters of a certain size, i. e. n molecules, say, are formed. These clusters can be characterized by a "generating function," which is a measure of the probability of the molecule being part of a cluster or not. This then leads to there being a mean number of monomers in the cluster, which is also a measure for the size. Therefore we may introduce

$$F_0(\theta) = \sum_n W_n(p)\theta^p, \tag{4.3}$$

where $W_n(p)$ defines the probability that any monomer has reacted with the cluster which already consists of n monomers. The variable θ is an auxiliary variable and p defines the simple reaction probability between monomers, so this equation is relatively general. Further progress can be made by introducing a more specific description of the molecules, i. e. the functionality (number of reaction arms) φ must be introduced. To do so we introduce the associated probability $F_1(p, \theta)$ that a monomer has not reacted with the cluster and is naturally associated with $(1 - p)$. Therefore we have

$$F_0(\theta) = (1 - p + p\theta F_1(p, \theta))^\varphi. \tag{4.4}$$

This equation is, in general, difficult to solve, except when a network which has only a tree-like structure and no loops is formed, (compare Fig. 4.1). This appears at the moment to be a strong restriction, but the inclusion of loops is associated with a more general theory which we will treat shortly. With this assumption we are in the position to write out an iterative process, which naturally accompanies the reaction. Thus we have

$$F_1(p, \theta) = (1 - p + p\theta F_1(p, \theta))^{\varphi-1}. \tag{4.5}$$

Such equations can be solved by iteration (under the tree structure assumption) and we quote the results:

$$F_0 = (1 - \varphi)^\varphi p \sum_{n=1}^\infty \frac{[(\varphi - 1)n]!}{[(\varphi - 2)n]!(n - 1)!} X^{n-1},$$

$$F_1 = (1 - \varphi)^\varphi \sum_{n=1}^\infty \frac{[(\varphi - 1)n]!}{[(\varphi - 2)n]!(n - 1)!} X^{n-1}, \tag{4.6}$$

where the variable $X = p\theta(1 - p)^{\varphi-2}$ has been introduced. One interesting quantity to look at is the molecular weight of the cluster. Therefore it is reasonable to make a power-law ansatz for the quantity F_1 in (4.5), i. e. $F_1(p, \theta) = w^{\varphi-1}$. Then we may rewrite (4.5) in the following more instructive form:

$$\frac{w - 1}{p\theta} = w^{\varphi-1} - \frac{1}{\theta}, \tag{4.7}$$

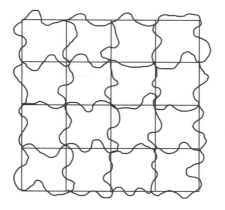

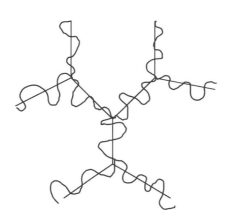

Fig. 4.1. Full loopy structure (lattice-like connectivity) in comparison with a tree-like structure. It is clear that their elastic properties will be different.

which indicates the critical point p_c. This can be seen from the following two cases:

- $p < 1/(\varphi-1)$: the molecular weight, associated with the value of w, is not diverging, the clusters stay finite, and no network results in the process.
- $p > 1/(\varphi-1)$: in this case the molecular weight diverges and one large cluster will emerge, i. e. a macroscopic network will form.

Therefore we can identify a "critical point" $p_c = 1/(\varphi-1)$ for network formation in this simple model. A critical reactive probability p_c is needed for the network (or better cluster) formation which depends on the functionality of the molecules. So far we have determined a critical point from just a simple model. This model, however, suffers from a serious drawback: The resulting structure is only tree-like; no further structural elements are allowed. Classical networks include loops and closed circles in their structure. Indeed, they define much of the elastic properties of the networks via the cycle rank [77, 78]. We will introduce a more general model for the network formation in the next section.

4.2 Percolation

The percolation process describes a more general process for cluster and network formation. In the simple Flory–Stockmayer model introduced above we saw that the main issue is a "connectivity transition" in which a liquid of zero elasticity of functional molecules undergoes a chemical reaction and connects molecules with each other. The resulting cluster is a solid with different properties, e.g. a finite elasticity. We have so far not learned anything about the exponents of the diverging viscosity of the growing elastic modulus.

Probably the simplest visualization of the percolation process is given by connecting bonds on a lattice [76, 183]. Imagine therefore a lattice in which the lattice points are connected by polymer chains (or simply by bonds) completely at random (see Fig. 4.2). Obviously at first only small clusters are formed. However, there might finally come a point, at which the connected cluster reaches from one end of the lattice to the other. The cluster can then be viewed as "infinite." The properties of this process can be summarized in a number of quantities starting with the size distribution of the clusters, which can be described by the following scaling approach:

$$n_s \propto s^{-\tau} f(s/s_0), \tag{4.8}$$

where the variable s corresponds to the size of the cluster and $f(s/s_0)$ is an as yet unknown scaling function that has to be determined later. Next we define an "order parameter" (in analogy to phase transitions), which is here associated with

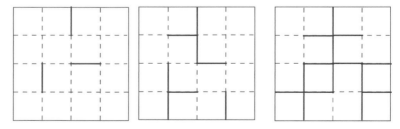

Fig. 4.2. Simple visualization of the percolation. Bonds are connected at random, then clusters of connected bonds appear. Eventually all the clusters have joined together to make a single entity.

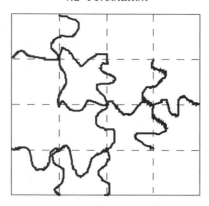

Fig. 4.3. The polymer version of the critical cluster. The bonds have been replaced by chains.

the probability that a connected lattice bond belongs to the infinite cluster:

$$P_\infty \propto \sum_s sn_s \propto (p - p_c)^\beta, \tag{4.9}$$

where p and p_c are as yet undefined. The exponent β is a so-called critical exponent, which should not depend on any details of the process, i. e. the form of the lattice and the length of the bonds, etc. It is supposed to be "universal." Note also, that the quantity P_∞ is obviously zero below p_c when no infinite cluster exists. Analogously the number of finite clusters, which is denoted by $G(p)$, can be described by a scaling law. It is given by the singular part of the sum over the cluster number:

$$G(p) \propto \sum_s n_s | \propto | p - p_c |^{2-\alpha}. \tag{4.10}$$

Of course, the percolation process is not limited to the case of connecting bonds. The bonds can be replaced by polymer chains, and then a critical network appears (see Fig. 4.3).

For the present purpose it is useful to define a reasonable measure of the molecular weight of the cluster, which corresponds to the degree of polymerization of the network. This is given by

$$S(p) \cong \bar{M}_w \propto \frac{\sum_s s^2 n_s}{\sum_s sn_s} \propto | p - p_c |^{-\gamma}. \tag{4.11}$$

Finally, for completeness we mention the correlation length ξ that defines the correlation between different bonds, which behaves as

$$\xi \propto | p - p_c |^{-\nu}. \tag{4.12}$$

So far we have defined a number of "critical exponents" which describe the behavior of the above quantities as we approach the percolation threshold. However, it turns out that these exponents are not independent but are connected by some scaling laws, which have their origin in the general theory of phase transitions. There are two essential scaling laws which we mention here without proof. The first is

$$\alpha + 2\beta + \gamma = 2, \tag{4.13}$$

and the second is

$$\nu d = 2 - \alpha = 2\beta + \gamma . \tag{4.14}$$

The latter scaling relation, which is called "the hyperscaling law," is essential since it connects universal scaling exponents to the space dimension d. The deep meaning of this formula becomes entirely clear when all the exponents from the Flory–Stockmyer model (here only quoted),

$$-\alpha = \beta = \gamma = 1,$$
$$\nu = \frac{1}{2}, \tag{4.15}$$
$$\tau = \frac{5}{2},$$

are inserted. The first scaling relation is easily satisfied, while the second is greatly violated, except in $d = 6$. Thus we should expect the Flory–Stockmayer theory to be valid only in six space dimensions.

How can we explain this? The explanation is, in fact, very simple since we have assumed in the solution of the corresponding equations that no loops are present and that the only structural element is a tree. The assumption of no loops can only be valid for large dimensions. In order to avoid having loops in a network-forming structure, a large dimensionality of the embedding space is necessary. Then there are always enough ways to avoid joining two ends without creating loops. Indeed, the tree formally corresponds to an infinite dimension. For the corresponding percolation problem six-dimensional lattices are sufficient to avoid loops.

The exponents of the percolation problem can in general only be determined numerically and are found to be $\alpha \approx 1.8$, $\nu \approx 0.88$, $\beta \approx 0.43$, and $t \approx 1.9$. Note that the value of the modulus exponent t in the Flory–Stockmayer theory is $t = 3$. So far we have only treated gel formation from small molecules. Rubbers are often formed from polymer chains, and we have to work out the main differences between the gelation and vulcanization processes. Note that t depends on the nature of the model, i.e. $t \approx 1.9$ results for scalar force constants, $t \approx 3.6$ for vectorial force constants (torsion).

4.3 Vulcanization

We have seen the remarkable difference between the exponents for the Flory–Stockmayer model and those for the percolation model. The main reason for this large difference lies in the fact that the Flory–Stockmayer model has no (structural) fluctuations, so that loops, dead ends, and blobs are forbidden as structural elements.

These elements appear naturally in any percolation model. However, tree-like structures become more probable as the number of dimensions increases. We have already indicated this when we considered the critical exponents together with the hyperscaling law which we could only satisfy at $d = 6$, the upper critical dimension. Of course, life in six dimensions is unimaginable, and vulcanizates exist in three dimensions. On the other hand, they can to a large extent be described by the classical theory of gelation. How can this be understood? It is due to the long preformed polymer chains [79].

To see this we again work in arbitrary space dimensions (although only $d = 3$ and $d = 2$ are accessible experimentally). In Section 4.2 we introduced a correlation length ξ that diverges with the reaction probability p in a critical way. Let us therefore consider a volume $V = \xi^d$, which contains the average number of monomers

$$n_t = c_0 \xi^d, \tag{4.16}$$

where c_0 is the average concentration. Now we assume that we have performed the reaction to slightly beyond the critical value p_c, so that a gel fraction already exists. This can be calculated as

$$n_{\text{gel}} = c_0 \xi^d P_\infty \propto c_0 \xi^d \left(\frac{p - p_c}{p_c} \right)^\beta. \tag{4.17}$$

Now consider the total (gel and finite) number of clusters and their fluctuations so that we have a balance equation of the form $\Delta n_{\text{gel}} + \Delta n_{\text{f}} = \Delta n_t \equiv 0$. Then the volume contains different types of monomers, which belong to finite and large clusters. The number of clusters is of the order of $n_{\text{f}}/S(p)$, and the relative fluctuation is of the order

$$\left[\frac{\Delta n_{\text{f}}}{n_{\text{f}}} \right]^2 \propto \frac{S(p)}{n_{\text{f}}}. \tag{4.18}$$

Further simplification is possible since we have assumed that we are working just above to the threshold, i. e. $p^+ \approx p_c$, which allows us to write $n_{\text{f}} \approx n_t$. The fluctuation then becomes

$$(\Delta n_{\text{gel}})^2 \sim n_t. \tag{4.19}$$

This yields with (4.16)

$$(\Delta n_{\text{gel}})^2 \propto S(p) c_0 \xi^d \tag{4.20}$$

The fluctuation of the gel fraction is then determined as

$$\left[\frac{\Delta n_{\text{gel}}}{n_{\text{gel}}}\right]^2 \propto \frac{S(p)}{c_0}\xi^{-d}P_\infty^{-2}(p)\,. \tag{4.21}$$

Now we can insert the $(p - p_{\text{c}})$-dependence for all quantities and find immediately

$$\left[\frac{\Delta n_{\text{gel}}}{n_{\text{gel}}}\right]^2 \propto (p - p_{\text{c}})^{\nu d - 2\beta - \gamma}, \tag{4.22}$$

which corresponds to the hyperscaling relation. However, there is even more in this formula. Using (4.21) we can distinguish the cases of percolation and vulcanization. In the vulcanization process, we have for the correlation length

$$\xi = bN^{1/2}\left(\frac{p - p_{\text{c}}}{p_{\text{c}}}\right)^{-\nu}, \tag{4.23}$$

while the gelation process corresponds to the chain length $N = 1$. Although there is only a difference in the prefactor, this makes a big difference between the two cases.

Let us ask therefore when the process corresponds formally to Flory–Stockmayer gelation. This is certainly the case if we have the mean field exponents (4.15). Therefore we use these exponents in (4.22) and see whether they make sense or not. Substituting, we find

$$\left[\frac{\Delta n_{\text{gel}}}{n_{\text{gel}}}\right]^2 \sim N^{1-d/2}\left(\frac{p_{\text{c}}}{p - p_{\text{c}}}\right)^{3-d/2}. \tag{4.24}$$

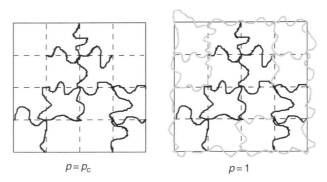

$p = p_{\text{c}}$ $p = 1$

Fig. 4.4. Complete conversion of a percolation process. The chains indicated in grey are now connected in comparison to the critical case $p = p_{\text{c}}$. When all lattice sites/bonds are occupied (grey chains) we have the case $p = 1$.

The gel point for vulcanization processes is now in general

$$p_{\mathrm{c}} = \frac{1}{N},\tag{4.25}$$

since the vulcanization reaction can occur at any place along the chain, which corresponds to large functionality, i. e. $\varphi = N$. Thus we have finally the criterion for the fluctuations:

$$\left[\frac{\Delta n_{\mathrm{gel}}}{n_{\mathrm{gel}}}\right]^2 \sim N^{-2}\left(\frac{1}{p - p_{\mathrm{c}}}\right)^{3-d/2}.\tag{4.26}$$

The fluctuation effect is very small when the molecular weight N is large. Therefore in three dimensions, $d = 3$, we can expect mean field behavior, except at a small distance close to the critical threshold $p_{\mathrm{c}} = 1/N$. In low dimensions (i. e. $d = 2$) the critical and non-classical behavior is more pronounced. This is well known from vulcanization of thin films, which corresponds to the case of $d = 2$. In the next chapter we will turn to the elastic behavior of a fully converted network. This happens when the reaction is complete, as depicted in Fig. 4.4, at which point $p = 1$, i. e. far above the percolation threshold.

5

The elastomer matrix

5.1 General remarks

The main goal of this chapter is to introduce a convenient view of the basic physics and elasticity of the rubber matrix. The easiest way to consider an elastic polymeric solid is as a crosslinked polymer melt. Polymer melts, however, already exhibit some properties of networks, at least on some time scales. This can be seen most beautifully by considering the storage modulus of a polymer melt.

The melt can be made a true solid by adding a reagent which joins each chain to a neighbor. For lightly crosslinked material there will be a few links per chain, but material can also be highly crosslinked [6]. Alternatively irradiation by gamma rays, X-rays, or by electrons will create crosslinks. There is ample evidence that polymers in melts are in random walk configurations, i.e. the molecule has a large choice of configurations and these differ by energies much less than the thermal energy $k_B T$. The kind of picture one has then is as in a computer simulation. The real difficulty is that rubbers are fundamentally three-dimensional and, unlike for crystals, two-dimensional pictures are not comprehensive. However, the reader can imagine a very kinky spaghetti-like mixture with permanent crosslinking bonds along the length. There is ample experimental evidence that perhaps 90% of the free energy of the material is entropic; see [6] for a general discussion and references.

In a network, however, the problem is that all the structural elements that make precise theories for melts difficult become frozen in. To see this we consider a polymer melt which consists of chains of a certain molecular weight. In this melt many structural elements are present, a few of which are shown in Fig. 5.1. In the ideal case, the chains are crosslinked to each other, but some chains have only one crosslink; these singly crosslinked chains have dangling ends and contribute to the density but not the elasticity. The main problem remains unsolved, which is to calculate the elasticity of the simplest representation of an elastic matrix.

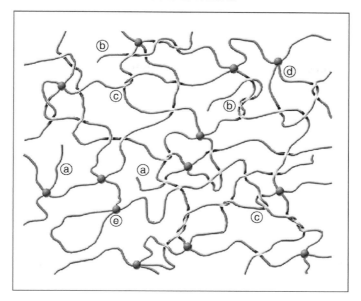

Fig. 5.1. Structural elements in a network: (a) dangling chain ends; (b) entangled chain ends; (c) trapped entanglements; (d) entangled loops; (e) wasted loops.

It is important to realize, that all the conformations of the chain described in this theory are purely entropic. The shape of the chain is driven purely by entropy. This means that the mean square chain radius fluctuates around its value given by (2.8). Indeed, this can be visualized by the following simplified formulation. The entropy of the chain is given mainly by the very beautiful Boltzmann formula $S = k_B \ln \Omega$ (which is written on his gravestone in the Zentralfriedhof in Vienna). To employ this fundamental equation for the present problem we write it in the form

$$S(\mathbf{R}) = k_B \ln P(\mathbf{R}), \tag{5.1}$$

which immediately yields the expression for the entropy of a chain with an end-to-end separation $\mathbf{R}$:

$$S(\mathbf{R}) = -k_B \frac{3R^2}{2b^2 N} - \frac{3}{2} \ln \left(\frac{3}{2\pi N} \right). \tag{5.2}$$

It is convenient to transform this into the free energy $F = U - TS$, but since the change in the internal energy U with respect to the chain end-to-end distance is zero no additional contributions are to be expected. Note also that the last term in (5.2) does not depend on the end-to-end distance $\mathbf{R}$ and is thus irrelevant for conformational changes.

5.2 The Gaussian network

It turns out that this knowledge of the Gaussian chain is sufficient to formulate the simplest theory for an elastomeric network. To see this we consider the entropy (or free energy) for the Gaussian chain. If we pull on a Gaussian polymer chain the force needed to extend it is given by

$$\mathbf{f} = \frac{\partial F}{\partial \mathbf{R}} = k_B T \frac{3}{b^2 N} \mathbf{R}. \tag{5.3}$$

Although this is a very simple formula it contains some interesting features. First, it can be seen that it resembles Hooke's law, where the force is proportional to the extension. Second, the force increases with increasing temperature. This is unusual in the sense that for ordinary elasticity it is well known that the force decreases with temperature, since the elasticity reduces. The physical reason for this typical entropic elastic effect is that at higher temperatures normally more conformations are adopted. If, on the other hand, the chain ends are held at a fixed distance $\mathbf{R}$ apart, this restricts the number of conformations. The larger the distance, the larger is the restriction, and according to (5.2) it falls off exponentially. Thus the force has to increase.

So far we have seen that chain extension costs entropy, but so does chain compression. This is straightforward to see from the above equations. It can be seen that the free energy has a scaling of the form

$$F_{\text{stretch}} = k_B T \, g(x), \tag{5.4}$$

where $g(x)$ is a scaling function and $x = R^2/b^2 N$. Of course, the form of $g(x) (\simeq x^2)$ has to be quadratic, since only classical elasticity is considered and the natural form of the entropy does not leave room for anything else at this level. For the compression of the chain, we can then guess the contribution of the free energy. The cost of confining a chain whose natural size is $R \simeq bN^{1/2}$ to any other (smaller) size R can be estimated from the entropy penalty

$$F_{\text{conf}} = \frac{3}{2} k_B T \frac{b^2 N}{R^2}, \tag{5.5}$$

which is just the inverse of the stretching formula. Of course, the chain fluctuates around any mean position. Thus we can estimate the size of the chain from the total free energy

$$F = \frac{3}{2} k_B T \left(\frac{R^2}{b^2 N} + \frac{b^2 N}{R^2} \right). \tag{5.6}$$

The minimum of this free energy occurs at $R \simeq bN^{1/2}$. Note that we have omitted all the numerical factors in the last few formulae. This is often done in the framework of scaling.

We can already use this small amount of information to construct a theory for a Gaussian network. This requires first some assumptions:

- The chains that form the network are Gaussian. This assumption seems to be sufficient inasmuch as the interactions in the melt are screened, as seen before. Therefore we may use Gaussian chains to form the network.
- All crosslinks are formed instantaneously and all chains are crosslinked.
- The density of the network is well behaved.
- The deformations on nanoscopic scales are the same as those on macroscopic scales.

Under these assumptions we can use the single-chain deformation behavior to calculate a free energy to the lowest order for a network. To do so, we introduce a diagonal deformation matrix

$$\boldsymbol{\lambda} = \begin{pmatrix} \lambda_x & 0 & 0 \\ 0 & \lambda_y & 0 \\ 0 & 0 & \lambda_z \end{pmatrix} \tag{5.7}$$

of deformation ratios $\lambda_{x,y,z}$ for the three principal axes. Each of the deformation ratios describes the ratio of the final to the initial length of the rubber as well as the individual deformations of the chains on the nanoscopic length scales, i. e.

$$\mathbf{R}(\boldsymbol{\lambda})_i = \boldsymbol{\lambda} \cdot \mathbf{R}_i, \tag{5.8}$$

where $\mathbf{R}_i$ is the undeformed end-to-end distance of the ith chain in the network. Then the free energy of a deformed rubber which consists of N_c crosslinked chains can be described as

$$F(\{\boldsymbol{\lambda} \cdot \mathbf{R}_i\}) = k_B T \frac{3}{2b^2 N} \sum_{i=1}^{n_c} (\boldsymbol{\lambda} \cdot \mathbf{R}_i)^2 . \tag{5.9}$$

Although this formula looks very trivial it contains some very interesting physics. Note that the free energy as it stands so far still contains "microscopic" variables, i. e. the individual chain end-to-end distances $\mathbf{R}_i$. Of course, any macroscopic free energy, which defines, e.g., the force–extension relationship must not depend on microscopic variables at all, since most of them are not known exactly. Therefore the free energy in (5.9) must be averaged over a suitable distribution of the microscopic variables. Actually, since the free energy in (5.9) is defined thermodynamically as a logarithm of a partition function, we have to average a logarithm. This is the first sign of a new formulation of (non-Gibbsian) statistical mechanics. We will come back to this point later.

For the present we proceed simply by averaging the free energy by the distribution already known for Gaussian chains. Therefore we average (5.9) with the previously calculated Gaussian distribution to receive the experimentally relevant free energy

$$F(\lambda) = \tfrac{1}{2}k_B T n_c \lambda \cdot \lambda^T. \tag{5.10}$$

As we have chosen a simple diagonal form for the deformation matrix, we rewrite the last equation in its standard form:

$$F(\lambda) = \tfrac{1}{2}k_B T n_c \left(\lambda_x^2 + \lambda_y^2 + \lambda_z^2\right). \tag{5.11}$$

This free energy enables us to compute a simple equation of state which here is a force–extension relationship. To do so we remark first that crosslinked polymer materials (in the absence of any solvents) are typical examples of bodies with a Poisson ratio close to $1/2$, i. e. there is no volume change during the deformation process. Therefore we can, e.g., for a uniaxial deformation experiment, use the relation

$$\lambda_z = \lambda,$$

$$\lambda_x = \lambda_y = \lambda^{-1/2}, \tag{5.12}$$

which yields immediately the force–extension relation

$$f = n_c k_B T \left(\lambda - \frac{1}{\lambda^2}\right). \tag{5.13}$$

Note that we have so far only used the deformation of the chain. If we had taken into account the compression term in (5.5) and (5.6) and also the macroscopic free energy, the equation would look more complicated. We leave the precise calculation for the reader, but mention that there will appear additional terms in the free energy of the form

$$\frac{1}{\lambda_x^2} + \frac{1}{\lambda_y^2} + \frac{1}{\lambda_z^2}$$

and

$$\frac{\lambda_x^2}{\lambda_y^2 \lambda_z^2} + \mathrm{perm}(x, y, z).$$

These additional terms have two features. The additional terms appear in more general theories. In particular, the sum of the inverse squares of the deformation ratios has a central meaning in the mathematical theory of the deformation of elastic solids: it is called the second invariant (whereas the sum of squares represents the first invariant). In this mathematical theory it is postulated that the deformation free energy depends only on the three invariants

$$I_1 = \lambda_x^2 + \lambda_y^2 + \lambda_z^2, \tag{5.14}$$

$$I_2 = \frac{1}{\lambda_x^2} + \frac{1}{\lambda_y^2} + \frac{1}{\lambda_z^2}, \tag{5.15}$$

$$I_3 = \lambda_x \lambda_y \lambda_z. \tag{5.16}$$

On the other hand, a simple molecular model of chain compression yields more complicated forms for the deformation dependence.

Although this free energy of deformation is far too simple to account for real experimental results, it shows many features which are needed for a more refined theory. Of course, the free energy derived so far is based on enormous simplifications. The main problem is that all sorts of network defects have been neglected. Therefore we are going to include topological constraints and chain entanglements. Moreover, finite chain extensibility has not been taken into account. Here the first guess is that the chain is finite and can only be stretched out completely from $\sqrt{N}$ to N. This would give a maximum extension of $\lambda_{\max} = \sqrt{N}$, which is often very much larger than that measured in reality. Nevertheless the problem of finite extensibility (but merely Gaussian chains) can be solved exactly. The theory predicts a force–extension relationship which corresponds to the theory derived above for small extensions, with strong deviations for large extensions. The force becomes infinitely large for $\lambda \to \lambda_{\max}$, which is logical, since the elasticity is purely entropic and at the maximum deformation there are no conformations left to choose [6].

In the next section we extend the Gaussian theory with respect to some of these aspects.

5.3 Entanglements and the tube model: a material law

The tube model (see Fig. 5.2) is due to Edwards and de Gennes and is a relatively simple representation of the topological state of a crosslinked melt [62, 63]. The picture now suggests a quite naive view of the physics, since most of the behavior seems to be determined by the tube itself. The chains inside the tubes play a subordinate role. Therefore a new length scale, which is supposed to rule the physics, is introduced: the tube diameter, or in other words, the mean distance between entanglements.

(a)

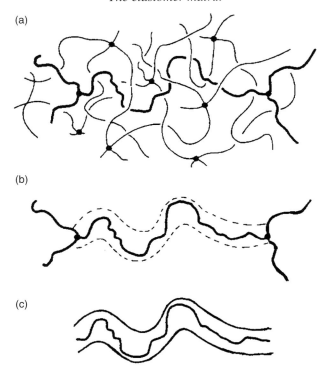

(b)

(c)

Fig. 5.2. Chains trapped in tubes as a simple mean field model for entangled states in networks: (a) a tagged chain in a polymer network; (b) the surrounding chains are replaced by a "tube"; (c) the pure tube model is restricted to the behavior of the chain trapped inside – in networks the chain cannot escape from the tube.

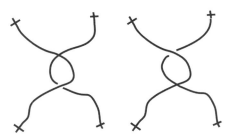

Fig. 5.3. Locally constrained chains. There is no way to transform the two chain conformations into one another.

However, a more detailed but even more symbolic view of entanglements can be given. Locally (at the edges of the tube) entanglements can be considered as local topological constraints as pictured in Fig. 5.3. The two conformations are not equivalent, since they cannot be transformed into each other. This is mainly due to excluded volume: the chains have a finite thickness. Therefore severe topological constraints exist. This means, that the number of configurations is some-how restricted, which has a strong influence on the mechanical behavior. On the other hand, the entanglements do not act as full crosslinks, since the constraint is not as severe as in fully fixed monomers. Therefore we may have more degrees of freedom from entanglements than from crosslinks. Thus we can expect that these topological constraints act at two extreme levels. At small deformations we may have restrictions of the sliding of the entanglements, whereas at larger deformations we have restrictions due to the additional tube geometry. These two points are the subject of the next two subsections, in which we mainly follow [7,42].

5.3.1 Entanglement sliding

The effects of the entanglements at low deformation can be mainly described by sliding. As mentioned before, we have already intuitively seen that an entangle-ment can be seen as a kind of "soft crosslink." Symbolically this can be drawn as in Fig. 5.4. There, the crosslinks at the end act as full constraints, while the entan-glement has been replaced by a ring, which can slide along the chains a certain distance a. This distance a is of the order of the mean distance between entan-glements (or in the most dilute case, i. e. fewer entanglements, the mean contour available between two consecutive crosslinks). There is an effect on the entropy when the ring can slide a distance a along the arclength of the polymers. Of course, this sliding can happen anywhere in the networks and for the moment we assume that we have N_s such sliplinks in the rubber. The number N_s of such sliplinks and the number of crosslinks are assumed to be known, so that we will end up with a two-parameter theory. The authors are aware of the crudeness of this picture, but feel that one should always start with the simplest model and this is the simplest

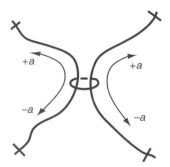

Fig. 5.4. The entanglement from Fig. 5.3 is replaced by a ring, which slides a certain distance $\pm a$ along the crosslinked chains.

model for studying configurational change in the presence of entanglement constraints. Later in this chapter (see Fig. 5.6) we will see how this can be discussed in a more general context when considering the physical behavior of the rubber. However, we will see later, that many experiments can indeed be described by this simple model.

The mathematical description of the statistical mechanics of this model is very complicated and is beyond the scope of this book. Instead, we provide a simplified version, along the lines given above in the theory of single Gaussian chain elasticity. To do so we start from the Gaussian probability equation (2.5) but allow for the slippage. Therefore we present a much simpler but nevertheless adequate derivation, after Wall–Flory [80]. It will turn out that this kind of approach is easier to generalize when one also needs to discuss inextensibility. We assume that the chain is Gaussian, with an end-to-end separation written in the form

$$P(\mathbf{R}, N) = \prod_{i=x,y,z} \int d\tau \chi(\tau) \left(\frac{3}{2\pi b^2 (N + \tau)}\right)^{3/2} \exp\left(-\frac{3R_i^2}{2b^2(N + \tau)}\right).$$

(5.17)

The additional variable τ describes the number of segments along which the slippage takes place. The distribution $\chi(\tau)$ describes the probability of the slippage. For simplicity, we replace the distribution by a simple rectangular probability, i. e.

$$\chi(\tau) = 1 \ \forall - a < \tau < +a.$$

(5.18)

Then we may write

$$P(\mathbf{R}, N) = \prod_{i=x,y,z} \frac{1}{2a} \int_{-a}^{+a} d\tau \left(\frac{3}{2\pi b^2 (N + \tau)}\right)^{3/2} \exp\left(-\frac{3R_i^2}{2b^2(N + \tau)}\right)$$

(5.19)

and follow the recipe to compute the free energy:

$$F = -k_B T \int d^3 R P(\mathbf{R}, N) \ln P(\boldsymbol{\lambda} \cdot \mathbf{R}).$$

(5.20)

After some calculations we find the free energy is given by

$$\frac{F}{k_B T} = \frac{1}{2} N_c \sum_{i=x,y,z} \lambda_i^2 + \frac{1}{2} N_s \sum_{i=x,y,z} \left\{ \frac{(1 + \eta)\lambda_i^2}{1 + \eta\lambda_i^2} + \ln(1 + \eta\lambda_i^2) \right\},$$

(5.21)

where we have added the contribution of N_c crosslinks and N_s sliplinks and η measures the effect of a macroscopically and is used as a parameter for the moment.

This result is consistent with the more complicated theory based on non-Gibbsian statistical physics by Ball *et al.* [81].

The free energy equation (5.21) is quite a simple result. Its main feature is a reduced macroscopic slip variable η, which is directly related to the mesoscopic variable a. Assuming the slip takes place between two neighboring crosslinks an estimation of η suggests values between 0 and 1. Two limiting cases should be discussed. First, if the slippage becomes zero ($\eta = 0$) the sliplink degenerates to a crosslink and contributes as a crosslink to the modulus. The free energy is then

$$\frac{F}{k_\mathrm{B}T} = \frac{1}{2}\,(N_\mathrm{c} + N_\mathrm{s}) \sum_{i=x,y,z} \lambda_i^2. \tag{5.22}$$

The other significant limit is given by infinite slippage, $\eta \to \infty$, although this appears unrealistic. Nevertheless, for large values of η the constraints are less severe. Such cases, i. e. large values of the slipping parameter η, can be crudely applied to swollen networks, in which the chain segments are pushed as far away from each other as possible. Then the free energy is given by

$$\frac{F}{k_\mathrm{B}T} = \frac{1}{2}N_\mathrm{c} \sum_{i=x,y,z} \lambda_i^2 + \frac{1}{2}N_\mathrm{s} \ln \prod_{i=x,y,z} \lambda_i \tag{5.23}$$

and the modulus depends only on the number of crosslinks. Note that $\prod_{i=x,y,z} \lambda_i = 1$, if the matrix volume remains constant.

Discussion of the physical behavior should now clarify major counterintuitive points. The intuitive picture of the behavior of a sliplink under stress is that it will respond by moving until it locks to another entanglement or crosslink and then behaves as a crosslink, perhaps of higher functionality. This means that a sliplink hardens, but a finite value of η in (5.21) suggests softening. This result is reasonable if one accounts for the increase in phase space during deformation for the slippage a. The slipping distance will be increased and more conformations will be accessible, so the link weakens. This can only be true, of course, if the deformation is not too large, i. e. it is not long before the polymers are drawn taut. We are going to investigate larger deformations in the following section.

5.3.2 Finite extensibility

So far all the theories are only applicable for small deformations where Gaussian statistics is valid. At larger values of deformation new effects occur. Consider a

single chain first. Kuhn and Grün [82] noted that finite extensibility of a single
chain is expressed as

$$\mathbf{f} = \frac{3k_B T}{b^2} \mathbf{R} \, \mathcal{L}\left(\frac{|\mathbf{R}|}{bN}\right), \tag{5.24}$$

where $\mathcal{L}(x)$ is the inverse Langevin function, which is well known in magnetism.
This function becomes singular when its argument reaches a value of 1, $x = 1$,
i. e. when the chain is fully stretched $|\mathbf{R}| = bN$. This result is obtained by adding
the constraint that the number of monomers is finite and fixed. For small single-
chain deformations $|\mathbf{R}| \ll bN$, this goes back to the Gaussian result, whilst in
the limit $|\mathbf{R}| \to bN$ the inverse Langevin function has a singularity, and $\mathbf{f}$ tends to
infinity. This is easily understood from the fact that at large extensions the entropy
decreases and when the polymer is stretched out there is no conformation left that
the polymer can occupy (compare with the book of Treloar [6] for an extended
discussion for consequences of (5.24)). The Kuhn theory suggests a maximum
value of the deformation of

$$\lambda_{\max} = \sqrt{N}, \tag{5.25}$$

which is a large number if this is applied to rubbers because it is the square root of
the number of segments between two crosslinks, and is never obtained in reality.
Therefore we look at what the tube model predicts.

The tube model's prediction of finite extensibility can be seen from geometrical
arguments. The answer to the problem lies in the definition of the primitive path.
The primitive path and the polymer are both random walks with the same end-
to-end distance. From Fig. 5.5 we see that the step length of the primitive path is
mainly determined by the end-to-end distance of the chain and the mean number
N_{pp} of entanglements per chain. Therefore, we recall the relations

$$\left(\frac{a}{b}\right)^2 = \frac{N}{N_{pp}},$$

$$\frac{a}{b} = \frac{L}{L_{pp}},$$

where L_{pp}, N_{pp} and a are the length, the number of segments and the step length
of the primitive path, respectively. The chains can now only be stretched until the
slack around the primitive path is used up. The amount of polymer slack is given by
the difference $L - L_{pp}$. According to Edwards and coworkers [83, 84] the primitive
path deforms as

$$L_{pp}(\lambda) = \left(\tfrac{1}{3}\boldsymbol{\lambda} \cdot \boldsymbol{\lambda}^T\right)^{1/2} L_{pp} = \left(\tfrac{1}{3}\sum_{i=x,y,z} \lambda_i^2\right)^{1/2} L_{pp}. \tag{5.26}$$

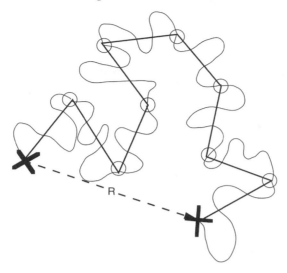

Fig. 5.5. Simple geometry of the primitive path. The primitive path and the chain have the same end-to-end distance. Therefore simple geometrical relations govern the length scales between them.

This is nothing but the Jacobian of a tangent vector along the contour under deformation. The polymer can only be deformed until all the slack is used up and the primitive path is taut. This occurs at $L - L_{pp}(\lambda) = 0$. For a uniaxial extension this gives a maximum value, λ_{max}, at

$$\lambda_{max} = \frac{a}{b}, \tag{5.27}$$

which is less than the $\sqrt{N}$ result predicted from Kuhn–Grün theory [82]. Typical values are 7–10, in reasonable agreement with experiment.

The consequences for the deformation dependence are slightly more complicated. We must show that with the tube model a singularity will occur at λ_{max}. It is well known [85, 86] that the joint probability distribution of the slack is given by

$$w(\Delta_i) = \left(\frac{1}{\Delta_0}\right)^{N_{pp}} \exp\left(-\frac{1}{\Delta_0} \sum_{i=1}^{N_{pp}} \Delta_i\right), \tag{5.28}$$

where the chain is modeled by N_{pp} primitive path steps and Δ_i describes the ith deviation from the primitive path of the real chain. The mean value Δ_0 is given by $\Delta_0 = a^2/b$, which indicates that the polymer slack itself behaves like a random walk between the primitive path steps. Therefore the total chain length can be written as

$$L = L_{pp} + \sum_{i=1}^{N_{pp}} \Delta_i. \tag{5.29}$$

The probability distribution of the primitive path is given by

$$P(L, L_{pp}) = \left\langle \delta \left(L - L_{pp} - \sum_i \Delta_i \right) \right\rangle, \tag{5.30}$$

where the average has to be taken over the slack probability $w(\Delta_i)$, see (5.28). The averaging can be carried out straightforwardly and produces a Gaussian distribution of the form

$$P(L, L_{pp}) = \frac{1}{\Delta_0 (L - \langle L_{pp} \rangle)^{1/2}} \exp \left(-\frac{(L - \langle L_{pp} \rangle)^2}{2\Delta_0 (L - \langle L_{pp} \rangle)} \right), \tag{5.31}$$

where $\langle L_{pp} \rangle$ is the mean length of the primitive path. Using the standard formula, we can calculate the deformation dependent free energy from the tube model:

$$\frac{F}{k_B T} = \frac{1}{2} N_c \left[\frac{(1-\alpha) \sum_{i=x,y,z} \lambda_i^2}{1 - (\alpha/3) \sum_{i=x,y,z} \lambda_i^2} + \ln \left(1 - (\alpha/3) \sum_{i=x,y,z} \lambda_i^2 \right) \right], \tag{5.32}$$

where the parameter α corresponds to the finite extensibility, $\alpha = (b/a)^2$. A similar result has been derived by Edwards [87], who showed that experimental data are in good agreement with this formula only at large deformations.

5.3.3 Tube and sliplinks

Finally we must seek a model that is valid for small and large deformations. In order to avoid calculations we need a simple mathematical tractable model of the tube or the primitive path into which the deformation can easily be brought. This has been carried out in detail in several papers so that for this very specialized calculation we refer the reader to [42]. Here we just quote the result:

$$\frac{F_s}{k_B T} = \frac{1}{2} N_s \left[\sum_{i=x,y,z} \left(\frac{(1-\alpha)(1+\eta)\lambda_i^2}{(1 - (\alpha/3) \sum_{i=x,y,z} \lambda_i^2)(1 + \eta\lambda_i^2)} + \ln(1 + \eta\lambda_i^2) \right) \right.$$

$$\left. + \ln \left(1 - (\alpha/3) \sum_{i=x,y,z} \lambda_i^2 \right) \right] \tag{5.33}$$

for the sliplink contribution, and

$$\frac{F_c}{k_B T} = \frac{1}{2} N_c \left[\frac{(1-\alpha)\sum_{i=x,y,z} \lambda_i^2}{1 - (\alpha/3)\sum_{i=x,y,z} \lambda_i^2} + \ln\left(1 - (\alpha/3)\sum_{i=x,y,z} \lambda_i^2\right) \right] \quad (5.34)$$

as before for the pure crosslink contribution. The total free energy is then additive, i. e.

$$F = F_s + F_c. \quad (5.35)$$

Therefore we have produced a constitutive materials law which contains the main information on the crosslinks and the entanglements based on a molecular model. We have to discuss next whether this model provides a basis for satisfactory agreement with experimental data.

5.4 Experiments

We now consider the experimental results. There is an enormous quantity of data in the literature and here we will only give a few examples that we consider to be in the main stream of the discussion, mainly in the context of the tube model. Experiments should involve the following:

(1) Quasistatic stress–strain measurements to give the modulus and the shape of the stress–strain curve. These measurements indirectly give the derivative of the free energy with respect to deformation.
(2) Neutron scattering of networks to give the radius of gyration and the wave vector dependence. This corresponds to the probability distribution.
(3) Dynamic mechanical measurements of crosslinked melts to compare with the theoretical stress–strain curve.

Here we will restrict ourselves to only mechanical measurements in more detail and leave the neutron scattering for further reading.

5.4.1 The stress–strain relationship

The simplest and most convenient way to measure the modulus of a rubber is a stress–strain measurement in uniaxial extension. This gives most of the information desired for the modulus and the quantities connected with the free energy. For uniaxial extension, data are conveniently represented by the Mooney plot. This is a longstanding type of representation and follows directly from the mathematical theory of rubbers:

$$f = (\lambda - 1/\lambda^2)(2C_1 + 2C_2/\lambda). \quad (5.36)$$

Thus on plotting $f^* = f/(\lambda - 1/\lambda^2)$ against $1/\lambda$ one should find a straight line with a slope of $2C_2$ and an intercept of $2C_1$. The modulus is given by the sum of both constants. An exhaustive discussion can be found in Treloar's book [6], where the Mooney representation (5.36), in which f^* is plotted against $1/\lambda$, is considered in detail. In the small deformation regime, $\lambda < 2.5$, the curves follow the Mooney plot surprisingly well, whereas for larger deformations large deviations occur.

The pure Gaussian model predicts a horizontal line, which is never the case for real systems, but it must be appreciated that these are gigantic strains in comparison to those in normal solids. Often combined extension and compression plots are used but at around $\lambda = 1$ awkward experimental problems are present and the data are less precise there. At high strains the data show a strong upturn. This is where inextensibility and crystallinity come in and the rubber hardens. At low strains the rubber softens (compared with an imaginary Gaussian model).

Some of the models quoted in the text above have been tested by Gottlieb and Gaylord [88–90] in a series of papers to which the reader is referred for the details. These authors tested the models by Edwards, Gaylord, Graessley and Marrucci [87, 91–93] and the constrained fluctuation model of Flory and Erman (see the book by Mark and Erman [94] for a broad discussion). Their conclusion was that the Gaylord model and the Flory model provided the best fit to their experiments. We now compare tube-type models and restricted fluctuation models, because these are the two types that are widely used. In particular, a central prediction is that of the modulus.

In Flory-type models the modulus is always given between the two limits of the phantom network

$$G = N_c \left(1 - \frac{2}{\varphi}\right) k_B T, \qquad (5.37)$$

where φ is the crosslink functionality. In the affine Kuhn model the modulus becomes

$$G = N_c k_B T. \qquad (5.38)$$

Note that there is no additional term from entanglements. This is very different from tube-type models, where entanglements are treated as physical constraints and there a strong dependence of the modulus is expected. In a simple phenomenological model, Langley [95] suggested a general behavior expressed as

$$G = G_c + G_e T_e, \qquad (5.39)$$

where the first term corresponds to the crosslink contribution and the second to the contribution by entanglements. T_e is the trapping factor and measures the strength of the constraint, i. e. the local coupling to the chains. Note that $T_e = 1$ corresponds to full coupling and the entanglements act as crosslinks. In our model, however, we can give a simplified expression for the modulus. This can be rewritten as

$$\frac{G}{k_{\mathrm{B}}T} = N_{\mathrm{c}}\frac{1 - 2\alpha^2 + \mathcal{O}(\alpha^4)}{(1 - 3\alpha^2)^2} + N_{\mathrm{s}}\frac{1 - 2\alpha^2 + 3\alpha^2\eta + 4\alpha^2\eta^2 + \mathcal{O}(\alpha^4)}{(1 + \eta^2)^2(1 - 3\alpha^2)^2}$$

$$\approx N_{\mathrm{c}} + N_{\mathrm{s}}\frac{1}{(1 + \eta)^2}, \quad (\text{for } \alpha \to 0) \tag{5.40}$$

which is exactly of the Langley form. The trapping factor is given by the slippage and has the correct limits. If $\eta = 0$ the sliplink is a crosslink and the two contributions are summed, and if $\eta \to \infty$ the sliplink is a weak constraint and G is given by the network contributions only. It is worthwhile mentioning, that the tube model has been investigated using biaxial stretching experiments and excellent agreement with the present theory was found [96].

5.4.2 The extended tube model of rubber elasticity

Basic assumptions

The extended, non-affine tube model of rubber elasticity is based on the assumption that the network chains in a highly entangled polymer network are heavily restricted in their fluctuations due to packing effects. This restriction is described by virtual tubes around the network chains that hinder the fluctuation. When the network elongates, these tubes deform non-affinely with a deformation exponent $v = 1/2$. The tube radius d_μ in spatial direction μ of the main axis system depends on the deformation ratio λ_μ as follows:

$$d_\mu = d_0\lambda_\mu^v, \tag{5.41}$$

where d_0 is the tube radius in the non-deformed state. The assumption of non-affine tube deformation ($v = 1/2$) is essential. It was initially derived based upon fundamental molecular statistical calculations [43,97,98] and later confirmed by applying scaling arguments [59,60,99]. Experimental evidence of non-affine tube deformations according to (5.41) is provided by neutron scattering of strained rubbers [100] and stress–strain measurements of swollen networks [105].

An extension of the non-affine tube model for applications up to large strains is obtained by considering that the network chains have a finite length and the stress in the network becomes infinitely large when the chain sections between two consecutive trapped entanglements are fully stretched. The trapping of chain entanglements by two crosslink points prevents the sliding of the chains across each other under extension, implying that the entanglement becomes an elastically effective network junction. The free energy density of the extended non-Gaussian tube model with non-affine tube deformation is then as follows [57–60, 102]:

$$
W_R \left(\varepsilon_\mu \right) = \frac{G_c}{2} \left\{ \frac{\left(\sum_{i=x,y,z}^{3} \lambda_\mu^2 - 3 \right) \left(1 - \frac{T_e}{n_e} \right)}{1 - \frac{T_e}{n_e} \left(\sum_{i=x,y,z}^{3} \lambda_\mu^2 - 3 \right)} + \ln \left[1 - \frac{T_e}{n_e} \left(\sum_{i=x,y,z}^{3} \lambda_\mu^2 - 3 \right) \right] \right\}
$$
$$
+ 2 G_e \left(\sum_{i=x,y,z}^{3} \lambda_\mu^{-1} - 3 \right). \tag{5.42}
$$

Note that this free energy density is related to the undeformed state in contrast to the situation in (5.21), (5.23), and (5.34). Here, n_e is the number of statistical chain segments between two successive entanglements and T_e is the trapping factor ($0 < T_e < 1$), which characterizes the portion of elastically active entanglements. The term in braces in (5.42) considers the constraints due to interchain junctions, with an elastic modulus G_c proportional to the density of network junctions. The second addend is the result of tube constraints, whereby G_e is proportional to the entanglement density μ_e of the rubber. The first addend also takes into account the finite chain extensibility by referring to (5.34) with $\alpha = T_e/n_e$ [58]. For the limiting case $n_e/T_e = \sum \lambda_\mu^2 - 3$, a singularity is obtained for W_R. This happens when the chains between consecutive trapped entanglements are fully stretched. It makes clear that the approach in (5.42) characterizes trapped entanglements as some kind of physical crosslinks (sliplinks) that dominate the extensibility of the network due to the larger number of entanglements as compared to chemical crosslinks. In the limit $n_e \rightarrow \infty$ the original Gaussian formulation of the non-affine tube model, derived by Heinrich and coworkers [43,97,98] for infinite long chains, is recovered.

The trapping factor T_e increases as the crosslink density increases, whereas n_e and G_e – as terms that are specific to the polymer – are to a great extent independent of crosslink density. For the crosslink and tube constraint moduli, the following relations to molecular network parameters hold:

$$
G_c = A_c \nu_{\mathrm{mech}} k_B T \tag{5.43}
$$
$$
G_e = \frac{\rho N_A b^2 k_B T}{4\sqrt{6} M_s d_0^2} = \frac{1}{\sqrt{6}} \nu_e k_B T. \tag{5.44}
$$

Here, ν_{mech} is the mechanically effective chain density, ν_e is the density of chains between consecutive entanglements, A_c is a microstructure factor, N_A is Avogadro's number, ρ is mass density, M_s and b are molar mass and length of statistical segments, respectively, k_B is the Boltzmann constant, and T is the absolute temperature.

The microstructure factor A_c takes into account the fluctuation of crosslinks: $A_c = 1$ for total suppression of crosslink fluctuations and $A_c = 1/2$ for freely fluctuating tetra-functional crosslinks (phantom networks). For a given fluctuation radius d_c of the crosslinks, it can be expressed by the error function $\mathrm{erf}(x)$ as follows [103, 104]:

$$A_c = \frac{1}{2} + \frac{1}{\sqrt{\pi}} \left(\frac{K_c \exp\left(-K_c^2\right)}{\mathrm{erf}\left(K_c\right)} \right) \tag{5.45}$$

with

$$K_c = \sqrt{6} \frac{d_c}{\langle R_c^2 \rangle^{1/2}}. \tag{5.46}$$

$\langle R_c^2 \rangle = Lb$ is the average end-to-end distance of intercrosslink chains in the undeformed state. For a derivation of (5.45) see also [101].

Applications for non-ideal networks

The tube model considered so far applies to a network structure with monodisperse chains that all contribute an equal amount to the elasticity properties. For such an ideal network with tetra-functional crosslinks the mechanically effective chain density ν_{mech} equals twice the density of crosslinks μ_c ($\nu_{\mathrm{mech}} = 2\mu_c = \nu_c$). In the case of non-ideal networks, as depicted in Fig. 5.1, the presence of defects like dangling chain ends, trapped entanglements, closed loop structures, and a polydisperse chain length distribution leads to significant deviations of ν_{mech} from its ideal value ν_c, i.e. the intercrosslink value. In the literature different estimates have been proposed to account for the defects of real networks [6, 57–60, 95, 101, 102, 105–111]. The effect of dangling chain ends is often considered according to an approach due to Mullins [106]:

$$\nu_{\mathrm{mech}} = \nu_c - \frac{\rho N_A}{M_n} \tag{5.47}$$

This involves the molecular weight M_n of the polymer chains before crosslinking. The correction term in (5.47) corresponds to the consideration of a gel point $\nu_c^* = 2\mu_c^* = \rho N_A / M_n$ of the network, i.e. the number of crosslinks necessary to connect all the chains of the initial uncrosslinked system to form a gel.

The additional influence of trapped entanglements on the density of mechanically effective chains was considered in a semi-empirical manner by Mullins as well [106]. By referring to the different fluctuation behavior of crosslinks and entanglements it can be expressed as follows [108–110]:

$$\nu_{\text{mech}} = \left(\nu_c - \nu_c^* \right) + \frac{A_e}{A_c} \nu_e T_e. \tag{5.48}$$

Here, T_e is the trapping factor of entanglements $(0 < T_e < 1)$, which is also termed the Langley trapping factor [95]. The quantity ν_e denotes the density of chains between consecutive entanglements which varies with the inverse of the squared tube radius d_0.

Equation (5.48) considers the combined effect of chain ends and trapped entanglements on the mechanically effective chain density. A_e is the microstructure factor of trapped entanglements which takes into account the fluctuations of the trapped entanglements. It can be determined from the fluctuation radius d_0 of entanglements in a similar procedure to that used for the crosslink estimation in (5.45) and (5.46) [101]. Due to the high mobility of trapped entanglements one can conclude that d_0 is significantly larger than d_c. Accordingly, the microstructure factor of trapped entanglements can be assumed to be well approximated by the value $A_e = 1/2$, which is the limiting value of (5.45) for large fluctuation radii. The microstructure factor A_c is now related to the mean end-to-end distance of all junctions, i. e. crosslinks and trapped entanglements. Hence, the average $< \cdots >$ in (5.46) has to be taken over both types of chains, which yields $A_c \approx 0.67$ independently of the crosslinking density [101].

From (5.42) the nominal stress $\sigma_{R,\mu}$ that relates the force F_μ in spatial direction μ to the initial cross-section $A_{0,\mu}$ is found by differentiation, $\sigma_{R,\mu} = \partial W_R / \partial \lambda_\mu$. For uniaxial extensions of unfilled rubbers with $\lambda_1 = \lambda, \lambda_2 = \lambda_3 = \lambda^{-1/2}$ the following relation can be derived:

$$\sigma_{R,1} = G_c \left(\lambda - \lambda^{-2} \right) \left\{ \frac{1 - \dfrac{T_e}{n_e}}{1 - \dfrac{T_e}{n_e} \left(\lambda^2 + \dfrac{2}{\lambda} - 3 \right)^2} - \frac{\dfrac{T_e}{n_e}}{1 - \dfrac{T_e}{n_e} \left(\lambda^2 + \dfrac{2}{\lambda} - 3 \right)} \right\}$$
$$+ 2 G_e \left(\lambda^{-1/2} - \lambda^{-2} \right). \tag{5.49}$$

For equi-biaxial extensions with $\lambda_1 = \lambda_2 = \lambda, \lambda_3 = \lambda^{-2}$ one finds for the nominal stress:

$$\sigma_{R,1} = G_c \left(\lambda - \lambda^{-5} \right) \left\{ \frac{1 - \dfrac{T_e}{n_e}}{1 - \dfrac{T_e}{n_e} \left(2\lambda^2 + \lambda^{-4} - 3 \right)^2} - \frac{\dfrac{T_e}{n_e}}{1 - \dfrac{T_e}{n_e} \left(2\lambda^2 + \lambda^{-4} - 3 \right)} \right\}$$
$$+ 2 G_e \left(\lambda - \lambda^{-2} \right). \tag{5.50}$$

In the case of a pure-shear deformation with $\lambda_1 = \lambda, \lambda_2 = 1$, and $\lambda_3 = \lambda^{-1}$ one obtains:

$$\sigma_{R,1} = G_c \left(\lambda - \lambda^{-3} \right) \left\{ \frac{1 - \dfrac{T_e}{n_e}}{1 - \dfrac{T_e}{n_e} \left(\lambda^2 + \lambda^{-2} - 2 \right)^2} - \frac{\dfrac{T_e}{n_e}}{1 - \dfrac{T_e}{n_e} \left(\lambda^2 + \lambda^{-2} - 2 \right)} \right\}$$
$$+ 2G_e \left(1 - \lambda^{-2} \right). \tag{5.51}$$

In the limit of small strains, $\lambda = (1 + \varepsilon) \to 1$ or $\varepsilon \to 0$, and by assuming the Gaussian limit $n_e \to \infty$ one obtains Young's modulus E_0 from a Taylor expansion of (5.49):

$$E_0 = \lim_{\varepsilon \to 0} \frac{\sigma_{R,1}}{\varepsilon} = 3 \left(G_c + G_e \right). \tag{5.52}$$

This shows that the Young modulus is strongly influenced by the deformation of the tubes, since the crosslink and topological constraint terms, G_c and G_e, contribute equal amounts.

5.4.3 Testing of the model

By fitting experimental data to (5.49)–(5.51) for the three indicated deformation modes, the model parameters G_c, G_e, and n_e/T_e, of unfilled polymer networks, can be determined. The validity of the concept can be tested if the estimated fitting parameters for the different deformation modes are compared. A "plausibility criterion" for the proposed model is formulated by demanding that all deformation modes can be described by a single set of network parameters. The result of this plausibility test is depicted in Fig. 5.6, where stress–strain data of unfilled natural rubber (NR) samples obtained at room temperature are shown for the three different deformation modes considered above. Obviously, the material parameters found from the fit to the uniaxial data provide a rather good prediction for the two other modes. The observed deviations are within the range of experimental error.

We point out that the material parameter G_e can, in principle, be determined more precisely by means of equi-biaxial measurements than by uniaxial measurements. This is due to the fact that the first addend of the G_e-term in (5.50) increases linearly with λ. This behavior results from the high lateral contraction on the equi-biaxial extension ($\lambda_3 = \lambda^{-2}$). It assumes a close dependency of the equi-biaxial stress on the tube constraint modulus, since G_c and G_e contribute nearly equally to stress at small and large extensions. For the uniaxial extensions described in (5.49) this is not the case. Here, the tube constraints lead to significant effects only for small extensions, since the G_e-term in (5.46) approaches zero as the λ value increases.

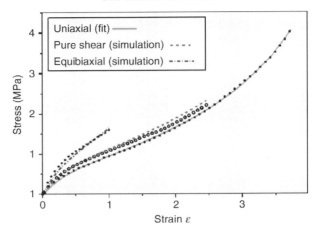

Fig. 5.6. Quasistatic stress–strain data (symbols) and simulation curves ((5.49)–(5.51)) of an unfilled NR for three deformation modes. The model parameters are found from a fit to the uniaxial data ($G_c = 0.43$ MPa, $G_e = 0.20$ MPa, $n_e/T_e = 68$). From [138].

Nevertheless, the experiments can be carried out more easily in the uniaxial case, and as a result, more reliable experimental data can be obtained.

For practical applications, the parameter G_e can also be determined from the value of the plateau modulus G_N, since the relationship $G_e \approx \frac{1}{2}G_N$ applies in accordance with the tube model. This implies that the parameter G_e is not necessarily a fit parameter but rather that it is specified by the microstructure of the rubber. We point out that the value $G_e = 0.2$ MPa obtained in Fig. 5.6 is in fair agreement with the above relation, since $G_N \approx 0.58$ MPa is found for uncrosslinked NR melts [112].

A further experimental test of the extended tube model of rubber elasticity focuses on the variation of the network parameters with the preparation conditions of the samples. Figures 5.7(a) and 5.8(a) show uniaxial stress–strain data measured at 100 °C for various crosslinked samples prepared from a high molar mass NR melt ($M_n = 248\ 000$) and a (mechanically treated) low molar mass NR melt ($M_n = 65\ 000$), respectively. The data were obtained for different amounts of crosslinker (CBS/sulfur), as indicated. The ratio between the vulcanization accelerator N-cyclohexylbenzothiazol-2-sulfenamide (CBS) and sulfur was kept constant (CBS/sulfur $= 0.18$). The solid lines are fitted according to (5.49). The correlation coefficient of the fittings is large in all cases ($R^2 > 0.999$). The development of the three fitting parameters G_c, G_e, and n_e/T_e with increasing amounts of crosslinker is shown in Figs. 5.7(b) and 5.8(b), respectively. Obviously, starting from a gel point at a crosslinker concentration of about 0.25 phr and 0.6 phr sulfur, respectively, both

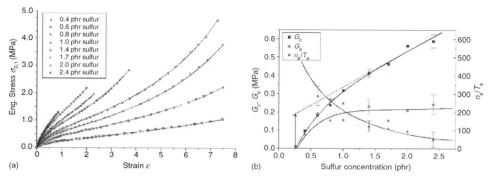

Fig. 5.7. Uniaxial stress–strain analysis of variously crosslinked, untreated NR-networks with high initial molar mass ($M_n = 249\,000$ g/mol): (a) Stress–strain data (symbols) and fittings (solid lines) according to (5.49). (b) Evaluation of fitting parameters with crosslinker concentration. The solid lines serve as guides to the eye. The dashed line corresponds to (5.43) and (5.48) with $T_e = 1$. From [102].

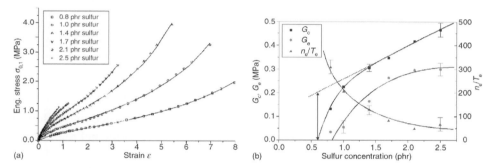

Fig. 5.8. Uniaxial stress–strain analysis of variously crosslinked, mechanically treated NR-networks with low initial molar mass ($M_n = 65\,000$ g/mol): (a) Stress–strain data (symbols) and fittings (solid lines) according to (5.49). (b) Evaluation of fitting parameters with sulfur concentration. The solid lines serve as guides to the eye. The dashed line corresponds to (5.43) and (5.48) with $T_e = 1$. From [102].

moduli G_c and G_e increase while the finite extensibility parameter n_e/T_e decreases. The topological constraint modulus G_e first increases and then approaches a plateau value located between 0.2 and 0.3 MPa in both cases. From (5.44) we expect this plateau value to equal almost half of the viscoelastic plateau modulus $G_N = \frac{4}{5} \nu_e k_B T \approx 0.58$ MPa of the NR melt [112]. The obtained plateau value of G_e is again in fair agreement with this expectation.

The behavior of the crosslink modulus G_c given by (5.43) can be understood if the form of (5.48) for ν_{mech} is considered. Therewith, the maximum entanglement contribution $\nu_e A_e / A_c$ corresponding to $T_e = 1$ can be evaluated if the limiting slope of G_c is extrapolated to the gel point (dashed lines in Figs. 5.7(b) and 5.8(b)).

Table 5.1. *Network parameters from uniaxial stress–strain, swelling, and nmr analysis of sulfur-cured NR networks at various crosslinker concentration and initial molar masses.* (*H1–H8: $M_n = 249000$ g/mol; L1–L6: $M_n = 65000$ g/mol*)

Sample	Sulfur (phr)	ν_{mech} (10^{-5} mol/cm^3)	ν_c(mech) (10^{-5} mol/cm^3)	ν_c(chem) (10^{-5} mol/cm^3)	ν_c(nmr) (10^{-5} mol/cm^3)	T_e (mech)	T_e (nmr)	E_0 (MPa)	$3(G_c + G_e)$ (MPa)
H1	0.4	6.0	1.9	3.7	2.6	0.36	0.24	0.86	0.53
H2	0.6	10.8	3.8	6.6	5.3	0.61	0.43	0.97	1.12
H3	0.8	14.1	6.2	8.8	6.3	0.69	0.52	1.13	1.17
H4	1.0	19.3	9.2	10.5	10.1	0.89	0.63	1.30	1.42
H5	1.4	25.0	13.9	13.8	11.5	0.97	0.70	1.69	1.92
H6	1.7	28.0	17.1	16.2	19.0	0.96	0.84	1.76	1.86
H7	2.0	34.0	21.0	18.3	17.5	1.14	0.84	2.12	2.31
H8	2.4	35.5	25.4	21.0	21.2	0.88	0.87	2.29	2.47
L1	0.8	8.0	1.8	4.5	2.3	0.52	0.41	0.67	0.50
L2	1.0	13.5	3.6	6.8	4.9	0.83	0.51	0.86	0.84
L3	1.4	18.4	7.0	10.3	7.8	0.95	0.64	1.25	1.41
L4	1.7	20.9	9.4	12.8	12.5	0.96	0.75	1.43	1.83
L5	2.1	25.1	12.8	16.0	14.1	1.03	1.15	1.78	2.13
L6	2.5	27.9	16.5	19.1	16.7	0.98	0.80	2.01	2.28

The deviations of the experimental G_c data from the limiting dashed lines allow an estimation of the trapping factors that increase from $T_e = 0$ at the gel points to its limiting values $T_e = 1$ at high crosslink concentrations.

The results of this evaluation procedure are summarized in Table 5.1, where, beside the values for ν_{mech} and ν_c (there denoted ν_c(mech)), the trapping factors T_e (there denoted T_e(mech)) are also listed. They are compared to equilibrium swelling results ν_c(chem) and nuclear magnetic resonance (nmr) results ν_c(nmr) and T_e(nmr), respectively, obtained with the same samples. More details of these evaluation procedures can be found in [102]. Analyzing the data of Table 5.1 one observes that, irrespective of sample series, the trapping factors T_e(mech) and T_e(nmr) both increase with increasing sulfur concentration. The two evaluation procedures give roughly the same results, but systematically smaller values are found from the nmr data. A comparison of the obtained intercrosslink chain densities shows fair agreement between ν_c(mech), ν_c(chem) and ν_c(nmr), though significant deviations occur in some cases. In particular, this indicates that the swelling equilibrium is mainly governed by the crosslinks and entanglement contributions are small (phantom chains). In the last two columns of Table 5.1, Young's modulus E_0, obtained at very small strains (below 5%), is compared to the sum $3(G_c + G_e)$, evaluated in the whole strain regime (Figs. 5.7(a) and 5.8(a)). It is obvious that the prediction of (5.52) is quite well fulfilled, confirming the applied non-Gaussian tube model of rubber elasticity.

In conclusion it has been shown that the initial chain length before crosslinking and the amount of crosslinker have a pronounced effect on the quasistatic mechanical properties of NR samples. The network parameters evaluated from uniaxial stress–strain data show fair agreement with the swelling and nmr analyses for the investigated NR sample series. In particular:

(1) The density of intercrosslink chains ν_c is found to increase almost linearly with the amount of sulfur if a critical concentration, referred to as the gel point, is exceeded (Table 5.1).

(2) For the mechanically degraded samples with low initial chain length the gel point is shifted to higher sulfur concentrations (≈ 0.6 phr) as compared to the untreated samples (≈ 0.25 phr) (Figs. 5.7 and 5.8).

(3) The crosslink efficiency (number of crosslinks formed per unit of sulfur) is somewhat reduced for the mechanically degraded systems (slope of the dashed lines in Figs. 5.7 and 5.8).

(4) The trapping factor T_e of entanglements increases with increasing amount of sulfur. It shows a weak dependency on initial chain length, only. For moderate amounts of sulfur just above the gel point it increases somewhat faster for the mechanically degraded systems (Table 5.1).

(5) The proposed tube model with non-affine tube deformations allows a reasonable description of quasistatic stress–strain data up to large strains. The predicted dependency, (5.52), of Young's modulus on crosslink and entanglement density is confirmed by the experimental data (Table 5.1).

In the literature more tests of the extended, non-affine tube model concerning stress–strain data of unfilled rubbers can be found that offer good agreement for various polymers and crosslinking systems [57–60, 102, 105, 108–110]. Further confirmation of the non-affine tube approach has been obtained in investigations considering mechanical stress–strain data and a transversal nmr relaxation analysis of differently prepared filler reinforced SBR networks [111]. The mechanical model used for these studies is treated in more detail in Chapter 10.

6

Polymers of larger connectivity: branched polymers and polymeric fractals

6.1 Preliminary remarks

So far we have discussed the behavior of linear polymer chains. Linear chains arise naturally from linear polymerization, a chemical reaction in which bifunctional (linear) monomers combine end to end to form a linear chain. We can associate an (internal) dimension to the linear chain by stretching it completely out and by coarse graining. Then we will have a one-dimensional object, i.e. the chain is of one-dimensional connectivity $D = 1$. This new dimension is related to the connectivity and is called the spectral dimension. If higher functional monomers are used for the polymerization process, branched polymers are generated, see Fig. 6.1. The challenge is how to describe branched polymers using methods similar to those we have used so far. Let us therefore try a simple generalization for the connectivity, in which we make use of the idea of higher-dimensional connectivity. Thus we try the more general Edwards Hamiltonian

$$\beta H = \frac{1}{2b^2} \int d^D \vec{x} \, (\nabla_{\vec{x}} \mathbf{R} \, (\vec{x}))^2 \; . \tag{6.1}$$

This is at first sight a strange object because it contains the internal and the spatial dimension. Here $\vec{x}$ are internal variables of dimensionality $\dim \vec{x} = D$. The boldface vector $\mathbf{R}$ describes the external variables with the dimension $\dim \mathbf{R} = d$, i.e. the dimension of embedding space. Naturally we must require $D \le d$. Let us try in the following to see if this analytical continuation makes sense. For simple visualization we show some examples for the analytic continuation of the spectral dimension in Fig. 6.2.

6.2 *D*-dimensionally connected polymers in a good solvent

The arguments above can be made more familiar if a common generalization of the Edwards Hamiltonian for linear polymers [62] is introduced for polymeric fractals [113] and D-dimensional manifolds in the following standard way:

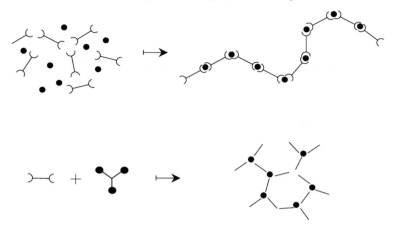

Fig. 6.1. Linear versus branched polymerization.

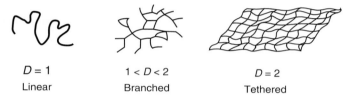

$D = 1$	$1 < D < 2$	$D = 2$
Linear	Branched	Tethered

Fig. 6.2. Different examples of the spectral dimension.

$$\beta H = \frac{1}{2b^2} \int d^D \vec{x} (\nabla_{\vec{x}} \mathbf{R})^2 - v \int d^D \vec{x} \int d^D \vec{x}' \delta^d (\mathbf{R}(\vec{x}) - \mathbf{R}(\vec{x}')) . \qquad (6.2)$$

D is the spectral dimension as noted earlier. Well-known special examples are $D = 1$ (linear polymers) and $D = 2$ (random tethered surfaces). The analytic continuation of D to non-integer values, i.e. $1 < D < 2$ corresponds to any polymeric fractal of arbitrary connectivity, $\vec{x}$ is a D-dimensional vector of the manifold embedded in d space dimensions described by the vector $\mathbf{R}$. The first term is the Gaussian connectivity and the second term the usual excluded volume interaction. In this book only objects with $D < 2$ are considered for convenience. The Hamiltonian (6.2) does not make sense for fractional values of D, without defining fractional differentials and integrals properly. In the scaling limit it can be used without problem. We restrict ourselves to this latter case.

Although we do not use the full Hamiltonian in this chapter, its introduction is helpful to derive its scaling properties [70, 114], especially for readers who are not familiar with the notation used below. The standard estimation uses the replacements $\mathbf{R}(\vec{x}) \rightarrow R$ and $|\vec{x}| \rightarrow N$. All integrations trivially yield a factor of N^D.

It must be remembered that N is not the total number of monomers but only the number of monomers in one given direction in spectral vector space.

Then Hamiltonians such as that given in (6.2) can be transformed easily into a Flory free energy by dimensional analysis [114]:

$$F = \frac{R^2}{b^2 N^{2-D}} + \frac{b^3 N^{2D}}{R^d} \, , \tag{6.3}$$

where we approximated the excluded volume by $v \cong b^3$. By substituting $M = N^D$ it is easy to show that (6.3) transforms into the well-known Flory form of polymeric fractals [113, 115, 116] with the Gaussian fractal dimension $d_f = 2D/(2 - D)$, which recovers cases of linear polymers ($D = 1$), randomly branched polymers ($D = 4/3$), and tethered membranes ($D = 2$). The case $D = 2$ corresponds to an infinite fractal dimension, which comes from the logarithmic size growth, i.e., $R \sim \log N$. The standard result is obtained by minimizing (6.3). This yields the usual Flory exponent for the size of the polymer in the swollen (crumbled) state, which is found to be [117]

$$v = \frac{D + 2}{d + 2} \, . \tag{6.4}$$

To avoid misunderstanding at this early stage, it has to be mentioned that this exponent accounts for the linear (chemical) size N and not for the total mass M. By simple arguments it is easily found that the corresponding fractal dimension is given by $d_f = D/v$. In the following we restrict ourselves to three-dimensional ($d = 3$) embedding space.

6.3 *D*-dimensionally connected polymers between two parallel plates in a good solvent

As a first example the case of a D-dimensional polymeric manifold between two plates in a good solvent is studied. The situation is depicted in Fig. 6.3. The first attempt to solve the problem is to use Flory's theory. This is very simple and the Flory free energy (see (6.3)) for this problem is given by

$$F = R_\parallel^2 / b^2 N^{2-D} + b^3 N^{2D} / (H R_\parallel^2) \, .$$

H is the distance between the two parallel plates and $R_\parallel$ is the size of the polymer parallel to the plates. Minimizing the free energy with respect to the size $R_\parallel$ yields the desired result:

$$R_\parallel = b \left(\frac{b}{H} \right)^{1/4} N^{(2+D)/4} \, . \tag{6.5}$$

Fig. 6.3. A linear polymer chain between two parallel plates. The polymer chain can be immediately replaced by a chain of any connectivity. For simplicity, only the case $D = 1$ is shown. Reprinted from [65] with permission from Elsevier.

Note that the N dependence in (6.5) corresponds to a two-dimensional polymer with spectral dimension D. For $D = 1$ (and $N = M$) the correct exponent $\nu = 3/4$ is recovered [63, 66]. For polymer sheets, such as flexibly polymerized membranes ($D = 2$), the reasonable exponent $\nu = 1$ is obtained. This corresponds to the case in which the tethered membrane is flat between two very narrow parallel plates with undulation fluctuations of the order of the distance between the two parallel plates. Although reasonable limits are predicted using this Flory model, it is difficult to be sure about the validity of this result if it is not derived by a different method, such as scaling theory. This will be done in the next paragraph.

The scaling analysis can be done in close analogy to the case of linear chains. The radius for the chain between two plates can be written as

$$R_{\parallel} = R_{\mathrm{F}} f \left(\frac{R_{\mathrm{F}}}{H} \right) , \tag{6.6}$$

where $R_{\parallel}$ is again the extension of the chain parallel to the plate and H the distance between the two plates. R_{F} is the geometrically unconstrained Flory radius in a good solvent and is defined by the Flory exponent given above. The scaling function $f(x)$ has two limits. The first is when $x = R_f/H$ tends to zero, i. e. when the plates are very far apart, then $f(x) \to 1$. In the opposite limit, when the parallel plates are placed very close together, the two-dimensional configuration appears, which determines the exponent corresponding of the scaling function in the usual manner. This corresponds to the two-dimensional configuration of the D-polymer and is calculated by the Flory model above. The usual argumentation provides the same answer as derived in (6.5). A more appropriate form is

$$R_{\parallel} = H \left(\frac{b}{H} \right)^{5/4} N^{(2+D)/4} . \tag{6.7}$$

It is tempting to generalize de Gennes's blob picture to such D-dimensional polymers. To do this, assume the D-polymer between two plates behaves as a fractal made out of blobs of size H. Thus it is reasonable to assume that the size of the object is given by $R_\parallel = H n^{(2+D)/4D}$, where n is the total number of blobs. Note that the fractal dimension of the effectively two-dimensional object $d_f = (2+D)/4D$ has been used to account for the mass in the fractal. The number of blobs n can be calculated by determining the mass m (number of monomers) of the blob. In the blob the branched structure shows good solvent behavior, which yields: $m = (H/b)^{5D/(2+D)}$. Therefore the number of such blobs is given by $n = M/m$, where M is the total mass $M = N^D$ as before. Following de Gennes' argumentation, the result

$$R_\parallel = H \left(\frac{b}{H}\right)^{5/4} M^{(2-D)/4D} \tag{6.8}$$

is obtained, which is identical to the results obtained using Flory theory and the scaling approach. This almost trivial example shows that the blob picture can be used to construct the same physically reasonable results for branched chains as for linear chains. The point is to also use the information from the scaling in terms of the number N of monomers in the chemical path in addition to the mass scaling in the blobs.

The cross-check for all the results is to consider the filling fraction $f = b^3 N^D/(H R^2)$ [63]. Later, see (6.11), it turns out that this quantity is useful in other respects. For the polymeric manifold (or polymeric fractal) in polymer between two parallel plates f is given by

$$f = \left(\frac{b}{H}\right)^{7/4} n^{(D-2)/2} .$$

This result makes physical sense, and for $D = 1$ the classical polymer behavior is recovered [63,118]. For $D > 2$ the filling fraction becomes unphysically large. Trivially, a polymeric membrane can be compressed completely between very narrow plates, i.e. $H = \mathcal{O}(b)$. In this special case the filling fraction becomes independent of the molecular weight, as is intuitively clear (lower critical dimension). The example of the arbitrarily self-similarly branched polymer between two plates has been discussed in more detail to demonstrate how the blob picture and the scaling arguments can be generalized to branched polymers or arbitrary higher-connected polymeric objects.

6.4 D-dimensionally connected polymers in a cylindrical pore (good solvent)

Severe problems occur when such self similarly branched, non-linear objects are forced into cylindrical pores, in other words when the space available for the polymer is further restricted, see Fig. 6.4. The simple dimensional analysis above has to be modified in the usual sense that the d-dimensional Dirac

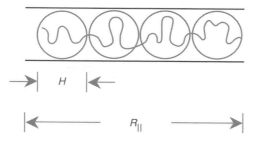

Fig. 6.4. A linear chain in a cylindrical pore. The chain is stretched lengthwise. Inside the blob the chain remains unperturbed by the pore size.

function becomes anisotropic and the lateral dimensions are given by the pore size. Thus we estimate the relevant excluded volume from (6.2) and (6.3) to be $\delta(\mathbf{R}(\vec{x}) - \mathbf{R}(\vec{x}')) \propto 1/(H^2 R_{\parallel})$, where H is now the pore diameter and $R_{\parallel}$ describes the chain extension along the cylindrical pore. Now we begin with a consideration of the Flory free energy for the manifold in the pore:

$$F = R_{\parallel}^2/N^{2-D} + vN^{2D}/(H^2 R_{\parallel}) \,. \tag{6.9}$$

Minimization immediately yields the result for the parallel exponent,

$$\nu_{\parallel} = \frac{D+2}{3} \,, \tag{6.10}$$

which agrees for $D = 1$ with the standard exponent [63], i. e. $\nu_{\parallel} = 1$, corresponding to the stretching of the linear chain along the pore. Obviously this exponent is ill defined whenever $D > 1$, i. e. whenever the polymers are of higher connectivity than linear ones. The linear dimension through the fractal or the membrane must not be larger than N itself, which corresponds to the completely stretched limit. It is now easy to understand that scaling and the blob model as presented for the slap cannot work for the case of a cylindrical pore. For example, if a scaling argument is considered which assumes good solvent behavior for large cylindrical pores and a linear (completely stretched) branched polymer for narrow pores, contradictions will show up, such as the filling fraction being unphysically large for all values for the spectral dimension $D > 1$. One way out of this difficulty is to postulate a minimum pore size through which the branched polymer is able to pass. Thus the minimum pore size can be defined by using the fact that the internal space of the branched polymer cannot be overstretched, i.e. ν cannot be larger than 1. In this case, the polymer fills the pore at maximum density, as depicted in Fig. 6.5. The minimum pore size is given by

$$H_{\min} \propto N^{(D-1)/2} \equiv M^{(D-1)/2D}. \tag{6.11}$$

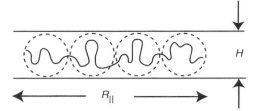

Fig. 6.5. A branched structure in a cylindrical pore. The connectivity is another limitation for the chain configuration in restricted geometries. This fact will become important in Chapters 8–10.

This result makes physical sense. The minimum pore size for linear polymers is independent of the molecular weight. Thus linear polymers ($D = 1$) find their way through even a very small pore if H_{min} is of the order of the Kuhn length, but with an extremely low probability and in a very long time. In branched polymers, with $D > 1$, another limitation is important: the connectivity. The larger the connectivity, the larger is the minimum pore size. It should therefore be possible to construct a porous medium that is able to separate a mixture of branched and linear molecules on the basis of their connectivity. This can be done by using an appropriate minimum pore size through which linear polymers can pass, while branched polymers cannot. For the construction of such a chromatograph dynamical aspects have to be taken into account, since the other selection constraint is the finite time taken to pass through a pore. Such aspects have been studied in detail by Gay [119].

The essential point to be made is that the minimum pore size is entirely defined by the spectral dimension and the molecular weight. Therefore the pore is able to select objects with respect to their connectivity, i. e. their spectral dimension. This possibility is called spectral chromatography in [119] – to distinguish it from classical chromatography, which selects only with respect to molecular weight. When, for example, a membrane is put into a pore that is so small that it cannot flatten out in the remaining space, it has either to crumple in a specific direction, if the pore is large enough, or to saturate (collapse) in the lateral direction in smaller pores. Whether the crumpled or collapsed case occurs is determined by the minimum pore size H_{min}.

It is obvious that in so-called Theta solutions, where the second virial coefficient (or the excluded volume defined by the pair interaction term) becomes zero, the effects of the geometrical restriction are less pronounced [120]. For a linear polymer the Θ-exponent is $\nu = 1/2$ which is less than the swollen exponent [63]. Thus the total size of a chain in a Θ-solvent is smaller than in a good solvent. For polymeric manifolds of larger connectivity this is also the case and it could be concluded that the limitation due to the pore size is less severe. This is not the case. The effect of such Theta solutions can be simply estimated by replacing the pair interaction term by the three-body virial coefficient, which is given by M^3/R^6. Then repeating a similar dimensional analysis, we find a new expression for H_{min}. It can be shown that the minimum pore size is given by the same value $H_{min} \propto N^{(D-1)/2}$ as above. We will not elaborate on this since it is well known that the Flory argumentation yields very bad values for the exponents ν in two dimensions, but exact values at the lower critical dimension for three-body interactions $d = 1$ and the upper critical dimension $d = 3$ [66]. Little is known about the validity of Flory values of branched structures and manifolds under Θ-conditions, and we will not discuss it further. The value for H_{min} under Θ-conditions is, however, an indication that the minimum pore size is determined by the connectivity only and not by thermodynamic conditions.

It was mentioned earlier that an interesting check on the consistency of the results is to calculate the internal concentration, or filling factor, $f = N^D/H^2 R_{\parallel}$. The filling fraction becomes independent of the molecular weight $M = N^D$ of the manifold at the point at which the pore size is minimum. This indicates that the pore size is a natural scale for D-dimensional polymers.

6.5 Melts of fractals in restricted geometries

The case of melts of fractals and branched polymers is also of interest. Melts of linear chains in small cylindrical pores have been studied by Brochard and de Gennes [121]. Again the case of $D = 1$ linear polymer melts in restricted geometries is simple. The generalization to branched polymers and polymers with higher connectivity is not as trivial and simple as it might look at first sight. From our previous discussion for polymer melts consisting of linear chains it is easy to see [116] that melts of branched polymers with a spectral dimension $D > 1$ must be divided into two classes. Whenever the connectivity is larger than a threshold value $D_s : D > D_s = 2d/(d + 2)$, the fractals do not interpenetrate in melts as linear chains do. In such systems the connectivity and space filling are too high. Instead of interpenetration the polymers saturate and form separate balls of their natural density, i.e. $R \propto N^{D/d}$. Using the same argumentation as before: take a melt of branched polymers and integrate the collective variables out to be left with one test fractal in the melt. The corresponding Hamiltonian is then given by

$$\beta H [\mathbf{R}_1(\vec{x})] = \frac{1}{2b^2} \int d^D \vec{x} \, (\nabla_{\vec{x}} \mathbf{R}(\vec{x}))^2$$
$$+ \frac{1}{2} \int d^D \vec{x} \int d^D \vec{x}' \sum_{\mathbf{k}} \frac{v}{1 + v S^0(\mathbf{k})} \exp\left\{i\mathbf{k} \left(\mathbf{R}(\vec{x}) - \mathbf{R}_1(\vec{x}')\right)\right\} ,$$

$$(6.12)$$

where the second term is the well-known effective potential. The Flory estimate again just uses the zero wave vector term and uses the structure factor $S^0 \sim C N^D (1 + 0(k^2))$, where C is again the concentration. This suggests that the melt value of the excluded volume parameter is of the order $v \propto N^{-D}$ [12]. The Flory free energy for melts is then

$$f = R_{\parallel}^2/N^{2-D} + b^3 N^D/R^d, \tag{6.13}$$

which leads formally to the D-independent melt exponent $v = 2/(d+2)$. This result makes no physical sense since it yields a fractal dimension larger than the space dimension: i.e. $d_f = D(d+2)/2$, which is larger than the space dimension d whenever the spectral dimension D is larger than $D_s = 2d/(2+d)$. For smaller connectivities the situation is very different and the result is that the polymers take their unperturbed size, which is given by $R \propto N^{(2-d)/2}$. This happens since for all cases the system is above the upper critical dimension. The upper critical dimension in the melt is given by

$$d_{uc} = 2D/(2-D), \tag{6.14}$$

which is of factor of 2 smaller than in the good-solvent case. This is due to excluded volume screening. The two different cases are now discussed in detail. For simplicity only the example of a cylindrical pore is considered. The discussion can be extended to slap very easily and straightforwardly. Below we consider two different cases: when the spectral dimension is larger than the critical value D_s, and when it is smaller.

(1) $D > D_s$ In this regime the melt of fractals is saturated. This means that, unlike linear polymers, polymers of higher connectivity do not interpenetrate each other since their connectivity is so large that this cannot happen. In one limit for such polymers in the pore, they form a row of balls each of size $R_s = bN^{D/3}$. The filling fraction for this situation is easily calculated and is given by

$$f = b^3 N^D/H^2 R_s = (b/H)^2 N^{D-D/3}. \tag{6.15}$$

The filling fraction cannot be larger than 1 and, therefore, the pore diameter is limited to values $H^* = bN^{D/3}$, which is the saturation radius of the polymer itself. This situation is sketched in Fig. 6.6. When the pore is smaller, each of the individual saturated fractals can be compressed further as their saturated radius of gyration can be elongated to form ellipses. Again the maximum parallel radius of the polymer is given by $R_{sm} = bN$, and for this case the filling fraction is

$$f = b^3 N^D/H^2 R_{sm} = (b/H)^2 N^{D-1} \tag{6.16}$$

and the limiting pore size is $H_{min} \propto N^{(D-1)/2}$, which is identical to that for good solutions and Θ-solutions. Again we find that the pore size does not depend on the thermodynamic state of the manifold. H_{min} is only determined by the molecular weight and the connectivity. To pass through this minimum pore size each of the

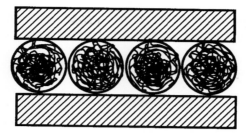

Fig. 6.6. A melt of highly connected polymers or manifolds in a small cylindrical pore. The connectivity is larger than $D = 6/5$. The manifolds cannot penetrate each other and they form separate balls. The pore diameter is identical to the radius of gyration in the saturated case. Reprinted from [65] with permission from Elsevier.

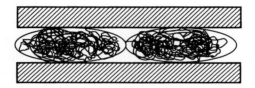

Fig. 6.7. Same in Fig. 6.6 but with a smaller pore diameter. The individual fractals can be compressed until they are stretched to their maximum radius of gyration $R^D \propto N$. Reprinted from [65] with permission from Elsevier.

saturated fractals has to be compressed by a factor of $\lambda = N^{-(D-3)/6}$, which for membranes is $\lambda = N^{1/6}$ and for randomly branched polymers is $\lambda = N^{1/9}$, as both cases belong to the class $D > D_s$. A visualization of this situation is shown schematically in Fig. 6.7.

(2) $D < D_s$ In this case the physical picture is very similar to that for linear polymers. As long as the connectivity is less than the critical $D_s = 6/5$ in three dimensions, the fractal takes its ideal Gaussian dimensions, i. e. $R_0 = bN^{(2-D)/2}$. This is because the upper critical dimension for the melt is always less than the dimension itself, i. e. in the present case $d = 3$. Therefore excluded volume forces are screened completely and the manifold behaves ideally. In addition this means that the manifolds can interpenetrate each other as the connectivity is very small. There are again two basic length scales for the pore size. The first one is given by the limit at which the manifolds just pass through without changing their shape. This can be read off from the filling fraction

$$f = b^3 N^D / H^2 R_0 = (b/H)^2 N^{(D-2)/2} . \tag{6.17}$$

The limiting value for the pore size is given by $D_0 = bN^{(2-D)/2}$. For a larger pore size the polymers in the melt are able to pass through without changing their shape. Smaller pore sizes are possible, if the manifolds stretch out. The maximum stretching is given by $R_m = bN$, and the limiting filling fraction predicts $D_{\min,\text{melt}} = aN^{(D-1)/2}$, which is again identical to the same limiting pore size as that in all other cases.

6.6 Once more the differences

In this chapter a scaling theory for arbitrarily connected polymeric manifolds under simple restriction has been presented. The main point is that larger connectivities give rise to severe limitations on the conformation and the behavior of such molecules in restricted geometries such as parallel plates or pores. The radius of gyration can be calculated for the case of two parallel plates using Flory theory, the blob model, and scaling theory. The blob model requires arbitrary spectral and fractal dimensions and to our knowledge has only been used for regularly branched molecules, such as star branched polymers [122].

These treatments can break down when the D-dimensional manifold is studied in a small cylindrical pore. Indeed, studying the behavior of the manifold in a cylinder has revealed the most serious restrictions, in contrast to linear polymers. Manifolds with larger spectral dimension than $D = 1$ do not pass through small pores without problems. A minimum pore size has to be assumed to obtain consistent results. Another important result is that this minimum pore size does not depend on the thermodynamic conditions of the manifold, such as good solvent, Θ solvent, or melt conditions. The minimum pore size is determined by the connectivity, i. e. the spectral dimension, only.

7

Reinforcing fillers

7.1 Fillers for the rubber industry

Reinforcement of elastomers by colloidal fillers, like carbon black or silica, plays an important role in the improvement of the mechanical properties of high-performance rubber materials. The reinforcing potential is mainly attributed to two effects: (i) the formation of a physically bonded flexible filler network and (ii) strong polymer–filler couplings. Both of these effects arise from a high surface activity and the specific surface of the filler particles [3, 8, 28, 123]. For a deeper understanding of structure–property relationships of filled rubbers it is necessary to consider the aggregate morphology and surface structure of fillers more closely. The present chapter is devoted to several technological applications of fillers in rubbers. In particular, we demonstrate how the physics of rubber nano-composites facilitates the understanding of how new generations of fillers, like silica (instead of carbon black), boost tire technologies giving simultaneously improved performance in rolling resistance and wet grip behavior.

Since the introduction of the Energy ® tire by Michelin, precipitated silica has proved (through partial or total substitution of carbon black) to be the filler of choice for the manufacture of high-performance pneumatic passenger car tires. The main reason is an improvement in the final compromise between the main interrelated tire performance parameters: it gives a significant improvement in tire performance in regard to rolling resistance, wet grip, and stopping distance for cars equipped with anti-lock braking system (ABS) steering [124]. These improved characteristics mean that silica-filled tread compounds are also the best available materials for winter performance [125]. It is noteworthy to observe that although a large number of studies have been reported in the literature, most of them were implemented around the original Michelin formulation disclosed by Rauline [126]; i. e. compounds made with silica as the main filler, a medium- or high-vinyl solution styrene–butadiene copolymer (S-SBR) as the main elastomer, and silicon-containing coupling agents.

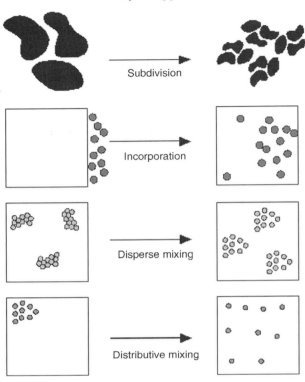

Fig. 7.1. Various steps in rubber filling.

Good mixing of silica compounds is a basic prerequisite to reach the required compound properties. Figure 7.1 illustrates – in a very general way – the various steps in rubber mixing [127]. The first elementary step of mixing is called "the subdivision," which is the breaking down of larger lumps of fillers to smaller ones, suitable for incorporation into a rubber matrix. This is followed by incorporation of powdered or liquid materials into the rubber to form a coherent mass. During this step the rubber penetrates into the void space of the agglomerates thereby replacing the trapped air. Dispersive mixing involves reduction of the size of agglomerates to their ultimate size, the aggregates; i. e. changing their physical state while at the same time distributing the particles formed. The dispersive mixing principle ensures that the particles are not only spread through the compound, but are also broken down into smaller entities. This leads to a greater reaction surface or contact area between the filler and the rubber. The breakdown of agglomerates in rubber requires a large amount of energy due to the high viscosity of the matrix. Equipment design and operating conditions must meet this criterion without reducing particle size. Although the detailed physical mechanisms associated with the breakdown

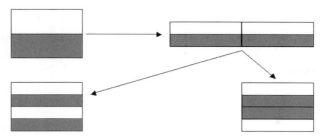

Fig. 7.2. Chaotic stacking.

process are not fully understood, it is generally agreed that to rupture the dispersed phase via interfacial hydrodynamic forces, the stress level in the continuous phase must exceed a critical value.

Distributive mixing, i.e. moving particles from one point to another, without changing their physical shape in order to increase randomness or entropy, is also called "extensive mixing." Distributive mixing can be compared to the mixing of a baker's dough with a roll, in which a portion of flour is incorporated in the dough. This type of mixing has several similarities to a square iteration process [128]. The mixing process is a chaotic process consisting of two different folding processes (Fig. 7.2). The folding can be done by stacking the layers in two different ways, and it is not predictable which way is used. As these processes are equivalent, it does not matter whether it is unknown which process actually occurs. It is interesting to note that both these processes are chaotic in nature. The flow system is defined as chaotic [129], if it satisfies any of the following criteria: (i) positive Lyapunov exponents in some region of the flow [130]; (ii) homoclinic or heteroclinic points (crossing streams) are present; (iii) Smale horseshoe functions [131] (folding and stretching) are present.

In the final step, viscosity reduction occurs by mechano-chemical breakdown of the polymer and its transformation into a more easily deformable and less elastic state.

7.2 Carbon black

7.2.1 Morphology of carbon black aggregates

Active fillers are commonly specified by different characteristic sizes. In the case of carbon black, the primary particle typically has cross-sectional dimensions of 5–100 nm. The size of the primary particles is commonly expressed in specific surface area/weight (m^2/g). Aggregates of multiple primary particles are formed by chemical and physical-chemical interactions: typically they have dimensions of 100–500 nm. The aggregate can be quantified by the number of primary particles and their geometrical arrangement. The term "structure" is commonly used to describe two arrangements: (i) low structure – linear arrangement; (ii) high structure – grape-like bundle. The aggregates are further condensed into agglomerates by van der Waals forces. Typical dimensions of agglomerates are in the order of magnitude of 1–40 μm. Agglomerates disintegrate during rubber mixing – to about the size of aggregates.

Carbon blacks for the rubber industry are produced in a variety of classes and types, depending on the performance required for the final product. In general, they consist of a randomly ramified composition of primary particles that are bonded together by strong sinter bridges. Significant effects of the different grades of carbon blacks in elastomer composites result from variations in the specific surface and/or "structure" of the primary aggregates [3,8]. The specific surface depends strongly on the size of the primary particles and differs from about 10 m^2/g for the very coarse blacks up to almost 200 m^2/g for the fine blacks. The "structure" of the primary aggregates describes the amount of void volume and is measured by, for example, oil (dibutylphthalate, DBP) absorption. It typically varies between 0.3 cm^3/g and 1.7 cm^3/g for furnace blacks.

The characteristic shape of carbon black aggregates is illustrated in Fig. 7.3, where transmission electron micrographs (TEMs) of five different grades of furnace blacks (N220, N326, N330, N347, N550) are shown. The variation in size of the

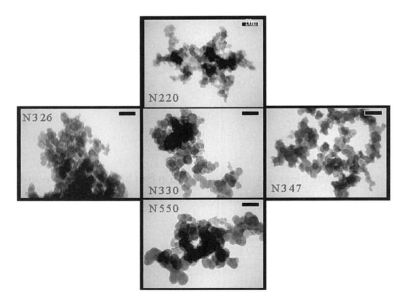

Fig. 7.3.   TEMs of five different grades of furnace blacks as indicated. The specific surface increases from top to bottom, the "structure" increases from left to right (bar length: 100 nm).

primary particles, increasing from top to bottom, is apparent. It implies a decline of the specific surface from 116 m^2/g for N220, 78 m^2/g for N326, 81 m^2/g for N330, 95 m^2/g for N347 to 41 m^2/g for N550. The "structure" or amount of specific void of the five grades increases from left to right and varies between 0.72 cm^3/g and 1.2 cm^3/g. Since the specific weight of carbon black is almost twice that of DBP, this corresponds to a factor 2 for the void volume as compared to the solid volume of the aggregates. This means that about 2/3 of the aggregate volume is empty space, i. e. the solid fraction ϕ_p of the primary aggregates is relatively small ($\phi_p \approx 0.33$). It is shown below that ϕ_p fulfills a scaling relation which involves the size and mass fractal dimension of the primary aggregates. Due to significant deviations of the solid fraction ϕ_p from 1, the filler volume fraction ϕ of carbon black in rubber composites has to be treated as an effective one in most applications, i. e. $\phi_{\text{eff}} = \phi/\phi_p$.

For a quantitative analysis of the structure of carbon blacks, such as those shown in Fig. 7.3, it is useful to consider the dependence of the solid volume V_p or the number of primary particles N_p per aggregate on aggregate size d. In the case of fractal objects one expects scaling behavior [132, 133]:

$$V_p \sim N_p \sim d^{d_f} . \tag{7.1}$$

The exponent d_f is called the mass fractal dimension or simply the fractal dimension. It characterizes the mass distribution in three-dimensional space and can vary between 1 and 3. This kind of fractal analysis of furnace blacks was performed, e.g., by Herd *et al.* [134] and Gerspacher and coworkers [135, 136]. The solid volume V_p of primary aggregates is normally determined (ASTM: 3849) from the cross-sectional area A and the perimeter P of the single carbon black aggregates by using a simple Euclidean relation [134]:

$$V_p = \frac{8A^2}{3P} . \tag{7.2}$$

However, it is not quite clear whether this relation can be applied for non-Euclidean ramified structures. Simulation results of carbon black formation under ballistic conditions by Meakin *et al.* [137] indicate that a scaling equation is fulfilled, approximately, between the number of particles N_p in a primary aggregate and the relative cross-sectional area A/A_p:

$$N_p \approx 1.51 \left(A/A_p \right)^{1.08} . \tag{7.3}$$

Here, A_p is the cross-sectional area of a single primary particle. The "structure" of carbon black aggregates can be quantified by considering the solid fraction ϕ_p.

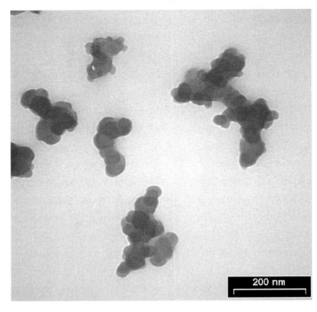

Fig. 7.4.   TEM of carbon black aggregates (N339) prepared from ready mixed
S-SBR composites with 40 phr filler (in-rubber state). From [138].

It is given by the ratio between the solid volume and the overall aggregate volume.
Then, with (7.1) one finds the following scaling relation with respect to the average
diameter d of the aggregates:

$$\phi_p = \frac{V_p}{(\pi/6)d^3} \sim d^{d_f-3} .\tag{7.4}$$

Depending on whether (7.2) or (7.3) is applied, significantly different values
for the mass fractal dimension are obtained. This discrepancy is demonstrated in
Figs. 7.4 and 7.5, by considering an example of a fractal analysis of primary carbon
black aggregates. Figure 7.4 shows a TEM of the furnace black N339 prepared
from a ready mixed composite of S-SBR after removing the unbound polymer.
This preparation procedure is indicated by the terminology "in-rubber state." It
was done by immersing the uncured composites for a week in a good solvent,
with the solvent being changed several times in order to remove the unbound
polymer. Afterwards the specimens were dispersed in a vibrator. The highly diluted
suspensions were then placed on a grid and carefully condensed. For the evaluation
of aggregate morphology (analogous to ASTM: 3849) roughly 500 particles of each
carbon black type were measured with respect to cross-sectional area A, perimeter
P, and diameter d. Double logarithmic plots of the solid volume V_p and the particle
number N_p, estimated from (7.2) and (7.3), vs aggregate diameter d are shown in

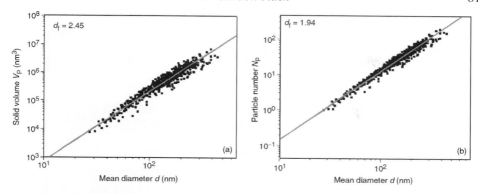

Fig. 7.5. (a) Fractal analysis according to (7.1) of primary carbon black aggregates (N339) prepared from S-SBR composites with 60 phr filler (in-rubber state). V_p is evaluated from (7.2). (b) Fractal analysis according to (7.1) of the same set of primary carbon black aggregates (N339) as shown in (a). The particle number N_p is evaluated using (7.3). From [138].

Figs. 7.5(a) and 7.5(b), respectively. The aggregate average diameter d is estimated as the mean value from 16 measurements on a single aggregate with a 15% variation in the angle of rotation. The fractal dimensions obtained differ significantly for the two evaluation procedures. From the slope of the two regression lines one finds $d_f = 2.45$ and $d_f = 1.94$, respectively.

In view of this discrepancy, we consider the conditions of primary aggregate growth during carbon black processing in some detail. Figure 7.6 shows a schematic representation of carbon black formation in a furnace reactor, where a jet of gas and oil is combusted and then quenched. As well as aggregate growth, resulting from the collision of neighboring aggregates, surface growth due to the deposition of carbon nuclei on the aggregates takes place during the formation of primary carbon black aggregates. The surface growth leads to the universal surface roughness, which is analyzed by the gas adsorption technique (see below). Obviously, the surface growth is also responsible for the strength of the primary aggregates, since it occurs in the contact range of the collided aggregates implying a strong bonding by sinter bridges (Fig. 7.6).

Due to the high temperature in the reactor, aggregate growth and surface growth both take place under ballistic conditions, i. e. the mean free path length of both growth mechanisms is large compared to the characteristic size of the resulting structures [137, 139, 140]. Therefore the trajectories of colliding aggregates (or nuclei) can considered to be linear. Numerical simulations of ballistic cluster–cluster aggregation yield a mass fractal dimension $d_f \approx 1.9 - 1.95$ [141–143]. This is comparable to the above TEM result $d_f \approx 1.94$ evaluated with (7.3). It means that the assumption of ballistic cluster–cluster aggregation during carbon

Reinforcing fillers

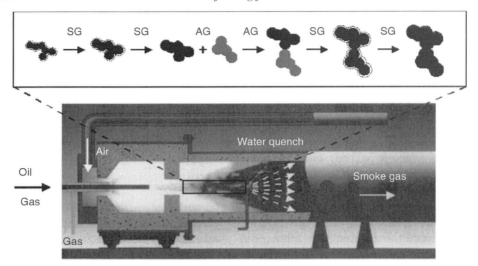

Fig. 7.6. Schematic view of carbon black processing in a furnace reactor. Primary aggregates are built by two simultaneous growth processes: (i) surface growth (SG) and (ii) aggregate growth (AG). From [138].

black processing, used in the derivation of (7.3), is confirmed by the TEM data for the relatively fine black N339. For more coarse blacks, with a typically small primary particle number, finite size effects can lead to a more compact morphology that differs from the scaling prediction of ballistic cluster aggregation. A further deviation can result from electrostatic repulsion effects due to the application of the processing agents (alkali metal ions) used to design the coarse blacks. Note that a relation similar to (7.3) was derived in the 1960s by Medalia and Heckman [144, 145]. The value $d_f = 1.94$ also agrees fairly well with other estimates obtained by e.g., electric force microscopy [147], TEM [146], and small-angle X-ray scattering (SAXS) [148–150].

Therefore, it appears likely that using the solid volume of primary aggregates evaluated from the two-dimensional cross-sectional area by (7.2) leads to an over-estimation of the mass fractal dimension. A more realistic estimate is obtained with (7.3). When (7.2) was used for the evaluation of the aggregate volume, the data obtained by Herd *et al.* [134] showed a successively increasing value of the mass fractal dimension from $d_f \approx 2.3$ to $d_f \approx 2.8$ with increasing grade number (or particle size) of the furnace blacks. As expected, they fit quite well to the above estimate $d_f \approx 2.45$ for the black N339. A summary of these data and a discussion including other fractal parameters can be found in [151].

In addition to TEM, a second technique that can be used to obtain information about the morphological arrangement of filler particles in elastomers is SAXS.

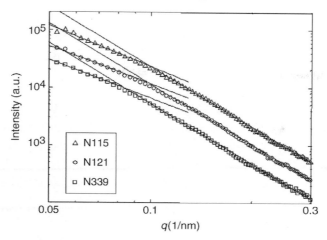

Fig. 7.7. SAXS data for the carbon black grades N115, N121 and N339 dispersed in NR.

Figure 7.7 shows the results of scattering investigations obtained for NR samples filled with different carbon black grades of varying specific surface. The concentration of carbon black is kept constant (46 phr). The double logarithmic plot in Fig. 7.7 demonstrates that in all cases two scaling regimes are obtained. For small values of the scattering vector q the slope is smaller than 3, indicating that the scattering is initiated by mass fractals. The mass fractal dimension d_f equals the negative value of the slope β ($d_f = -\beta$), and is found to vary between $d_f \approx 2.1$–2.4. This lies between the two values obtained above from the TEM data for the primary aggregates of N339. The lower cut-off length l_c is obtained from the cross-over points of Fig. 7.7 and increases somewhat with decreasing specific surface (50 nm $< l_c <$ 80 nm). Its value is of the order of the diameter of the primary particles, which seems reasonable since the primary particles represent the smallest units of the primary aggregates.

For large values of q the slope β in Fig. 7.7 is almost constant and larger than 3. This is due to a surface scattering by the filler particles. Accordingly, the surface fractal dimension $d_s = 6 - \beta$ is found to be almost independent of the specific surface ($d_s \approx 2.5$). This is evaluated in the length scale regime between roughly 60 nm and 20 nm. It is comparable to scattering results investigated in [150]. Note, however, that for smaller length scales of about 6 nm one again observes a crossover of the scattering intensity [148]. Hence, the surface fractal dimension $d_s \approx 2.5$ is related to the surface roughness on a mesoscopic length scale regime and does not reflect the surface roughness on atomic length scales, as obtained by the gas adsorption technique [152–154]. This will be considered more closely in the next section.

7.2.2 *Surface roughness of carbon blacks*

Different experimental techniques have been applied for the characterization of surface roughness of carbon blacks. As well as microscopic investigations, e.g., atomic force microscopy (AFM), that give an impressive but more qualitative picture, scattering techniques such as small-angle neutron scattering (SANS) [155] and SAXS [148–150], and gas adsorption techniques [152–154, 156–163] have been used for a fractal analysis of surface roughness. The results discussed in the literature appear somewhat contradictory, since almost flat surfaces with $d_s \approx 2$ [160, 161] and also rough surfaces with $2.2 < d_s < 2.6$ [148–150, 152–159] have both been found.

The reason for these discrepancies lies on the one hand in the restricted resolution of SANS and SAXS, since the scattering data can only be evaluated for wave vector $q < 1\,\mathrm{nm}^{-1}$ in most cases. This corresponds to length scales larger than about 6 nm, while the gas adsorption data typically were obtained at length scales smaller than 6 nm. More recent investigations by SAXS have gone down to smaller length scales with $q > 1\,\mathrm{nm}^{-1}$, where scattering from the graphitic layers at the carbon surface was observed. This means that the surface scattering was shielded by that of sheet-like structures [148]. On the other hand the discrepancies between the gas adsorption results arise primarily from the evaluation procedure of the effective cross-sections σ of the different gases if the yardstick method in the monolayer regime is used. The estimation of surface fractal dimensions in the multilayer regime is complicated by the fact that contributions of two different surface potentials have to be considered, those resulting from van der Waals and surface tension interactions. Depending on which of the two potentials dominates, remarkably different estimates of surface roughness are obtained. For that reason a proper analysis of these factors is necessary to obtain reliable results.

Here we will present results from two different evaluation procedures for the surface roughness of carbon blacks. In the monolayer regime we refer to the scaling behavior of the estimated Brunauer–Emmet–Teller (BET) [269] surface area with the size of adsorbed probe molecules (the yardstick method). On smooth flat surfaces the BET area is independent of the adsorbed probe or applied yardstick, while on rough surfaces it decreases with increasing probe (yardstick) size due to the inability of the large molecules to explore smaller cavities. This behavior is shown schematically in Fig. 7.8.

In the case of carbon black a power law behavior of the BET surface area with varying yardstick size is observed, indicating a self-similar structure of the carbon black surface. Double logarithmic "yardstick-plots" of the BET monolayer coverage N_m vs cross-section σ_ϱ of the probe molecules are shown in Fig. 7.9 for an original furnace black N220 and a graphitized (deactivated in an N_2 atmosphere at

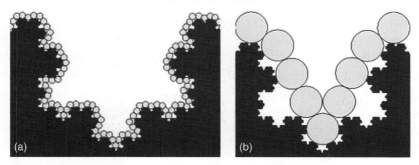

Fig. 7.8. Schematic presentation of a monolayer coverage of a fractal surface: (a) with small gas molecules; (b) with large gas molecules. From [138].

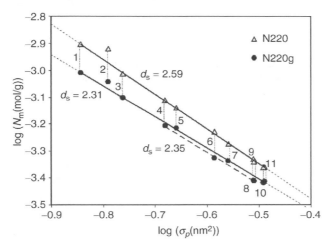

Fig. 7.9. Yardstick plot (equation (7.5)) of N220 ($\triangle$) and a graphitized N220g ($\bullet$) with adsorption cross-section σ_ρ determined from the bulk liquid density ϱ (equation (7.6)); 1 argon, 2 nitrogen, 3 methane, 4 ethene, 5 ethane, 6 propene, 7 propane, 8 isobutene, 9 n-butene, 10 iso-butane and 11 n-butane. The slopes yield $d_s \approx 2.6$ for N220 and $d_s \approx 2.3$ for N220g. Adsorption temperatures and densities ϱ are chosen according to the evaporation points of the gases at 1000 mbar. From [163].

$T = 2500\,^\circ\mathrm{C}$) sample N220g. The figure demonstrates that the roughness exponent or surface fractal dimension d_s differs for the two carbon black samples. By using the relation introduced by Mandelbrot [132, 133]:

$$N_m \sim \sigma_\varrho^{-d_s/2} \tag{7.5}$$

and neglecting the measurement points of nitrogen (N220 and N220g) as well as the measurement points of the alkenes (ethene, propene, isobutene and n-butene) one obtains from the slopes of the two regression lines of Fig. 7.9 a surface fractal dimension $d_s \approx 2.6$ for the N220 sample and $d_s \approx 2.3$ for the graphitized N220g sample. An extrapolation of the regression lines yields an intersection at an ultimate cross-section that corresponds to a yardstick length of about 1 nm, indicating that graphitization reduces the roughness of carbon black on small length scales, below 1 nm, only. Figure 7.9 clearly demonstrates that the reduction of BET surface area during graphitization is length scale (yardstick) dependent, proving that it is related to a change of surface morphology and is not, e.g., a result of reduced energetic surface activity.

An important point in the above evaluation of carbon black surface morphology is the correct estimation of the cross-section σ of the applied probe molecules. This is done by referring to the mass density ϱ of the probe molecules in the bulk liquid state, which are considered as spheres in a hexagonal close packing:

$$\sigma_\varrho = 1.091 \left(\frac{M}{N_A \varrho} \right)^{2/3} . \tag{7.6}$$

Here, M is the molar mass of the probe molecules and N_A is Avogadro's number. The crucial point now is the temperature dependence of ϱ which differs for the different probe molecules, mainly due to variations in the characteristic temperatures, e.g. the evaporation points.

We found that (7.6) can be applied without further corrections and high correlation coefficients of the "yardstick plots" in Figs. 7.9 and 7.10 are obtained only if the temperature during the adsorption experiments of a chemically similar, homological series of gases is chosen according to the same reference state, as defined in the framework of the theory of corresponding states. This is demonstrated in the "yardstick plots" of Fig. 7.10, showing that for the same carbon black (N220g) a different scaling factor is obtained within one series of gases if the adsorption temperatures are chosen with respect to different reference pressures, i.e. the evaporation temperatures at $p_0 = 10^3$ mbar and $p_0 = 10^4$ mbar, respectively.

A different scaling factor is also observed for each of the two different homological series of gases, i.e. the alkanes and alkenes. Nevertheless, the scaling exponent and hence the surface fractal dimension $d_s \approx 2.3$ is unaffected by the choice of the reference pressure or the adsorption gases applied. The alkanes methane, ethane, propane and isobutane are inert gases and roughly spherical in shape, but, due to their double bonds, the alkenes are not as inert as the alkanes and argon. Thus the interaction potential should be slightly different. The same holds for nitrogen which has a triple bond between its two atoms. As depicted in Fig. 7.9, the measurement points for the N220 sample, except for the measurement point of nitrogen, are on the straight line for a surface fractal dimension of $d_s \approx 2.6$. For the graphitized N220g the situation is different (Figs. 7.9 and 7.10). Here the points for the alkenes are all below the straight line for the alkanes and argon. But the points for the alkenes

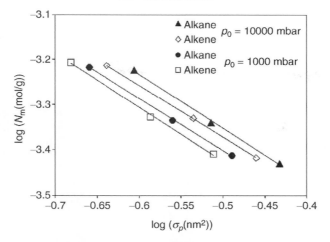

Fig. 7.10. Yardstick plots (equation (7.5)) of the graphitized carbon black N220g obtained with a series of alkenes (ethene, propene, isobutene) and alkanes (ethane, propane, isobutane). Adsorption temperatures are chosen as evaporation points at vapor pressures $p_0 \approx 10^3$ mbar and $p_0 \approx 10^4$ mbar of the condensed gases, respectively. From [138].

are also on a straight line which is parallel to the one for the alkanes and argon, indicating an identical value for the surface fractal dimension of $d_s \approx 2.3$ for the graphitized sample N220g. A discussion of the role of the interaction potential in the evaluation of yardstick plots from gas adsorption measurements can be found in [152, 163].

An alternative approach to the characterization of the surface morphology of carbon blacks by gas adsorption techniques is to consider the formation of a film of adsorbed molecules in the multilayer regime. In this case, the surface roughness is evaluated with respect to a fractal extension of the classical Frenkel–Halsey–Hill (FHH) theory, where, as well as the van der Waals surface potential, the vapor–liquid surface tension has to be taken into account [162, 164]. Then the Helmholtz free energy of the adsorbed film is given as the sum of the van der Waals attraction potential of all molecules in the film with all atoms in the adsorbent, the vapor–liquid surface free energy and the free energy of all molecules in the bulk liquid. This leads to the following relation between the number N of molecules absorbed and the relative pressure p/p_0 [162, 164]:

$$N \sim \left(\ln \frac{p_0}{p} \right)^{-\vartheta} \tag{7.7}$$

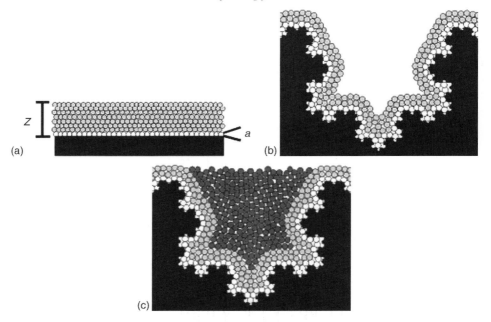

Fig. 7.11. Schematic view of the coverage of: (a) a smooth and (b), (c) a fractal surface according to the fractal FHH theory; o monolayer regime, FHH regime, ● CC regime; z: average film thickness, a: monolayer thickness. From [138].

with

$$\vartheta = \frac{3 - d_s}{3} \; , \; \text{FHH regime;} \tag{7.8}$$

$$\vartheta = 3 - d_s, \; \text{capillary condensation (CC) regime.} \tag{7.9}$$

The different exponents for the FHH and CC regimes consider the cases where adsorption is dominated by the van der Waals potential and the vapor–liquid surface tension, respectively. The two cases are shown schematically in Figs. 7.11(b) and 7.11(c), respectively. Note that in the CC regime a flat vapor–liquid surface is obtained due to a minimization of curvature by the surface tension. In contrast, in the FHH regime the vapor–liquid surface is curved, since it is located on equipotential lines of the van der Waals potential with constant distance to the adsorbent surface.

At low relative pressures p/p_0 or thin adsorbate films, adsorption is expected to be dominated by the van der Waals attraction of the adsorbed molecules to the solid which falls off with the third power of the distance to the surface (FHH regime, (7.8)). At higher relative pressures p/p_0 or for thick adsorbate films, the number N

of adsorbed particles is expected to be determined by the surface tension γ of the adsorbate–vapor interface (CC regime, (7.9)), because the corresponding surface potential falls off less rapidly, with the first power of the distance to the surface only. The cross-over length z_{crit} between the two regimes depends on the number density n_{p} of probe molecules in the liquid, the surface tension γ, the van der Waals interaction parameter α, and the surface fractal dimension d_s [162, 164]:

$$z_{\text{crit}} = \sqrt{\frac{\alpha n_{\text{p}}}{(d_s - 2)\, \gamma}}\,. \tag{7.10}$$

Note, that the cross-over length z_{crit} decreases with increasing surface fractal dimension d_s, implying that the FHH regime may not be observed on very rough surfaces. Then the film formation is governed by the surface tension γ on all length scales $z > a$ (compare Fig. 7.13).

The film thickness z is related to the surface relative coverage N/N_{m} and the mean thickness $a \approx 0.35$ nm of one layer of nitrogen molecules [165] according to the scaling law [132, 133]:

$$\frac{N}{N_{\text{m}}} = \left(\frac{z}{a}\right)^{3-d_s}. \tag{7.11}$$

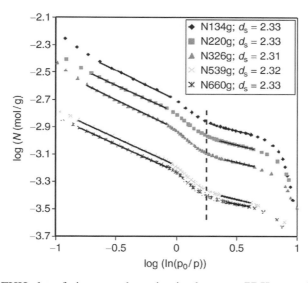

Fig. 7.12. FHH plot of nitrogen adsorption isotherms at 77 K on various graphitized furnace blacks, as indicated. The dashed line characterizes the transition between the FHH and CC regimes. The d_s-values, listed in the key, refer to the FHH regime at low pressures. From [138].

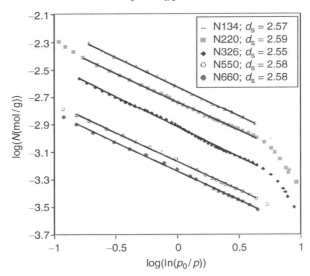

Fig. 7.13. FHH plots of nitrogen adsorption isotherms at 77 K of various furnace blacks, as indicated. The surface fractal dimension appears to be universal,
i. e. it varies between $d_s = 2.55$ and $d_s = 2.59$ for the depicted furnace blacks.
From [138].

N_m can be estimated from a classical BET plot and hence the film thickness z can
be obtained directly from N if the surface fractal dimension d_s is known.

So-called FHH plots of the nitrogen adsorption isotherms at 77 K of various
graphitized furnace blacks are shown in Fig. 7.12. The graphitized furnace blacks
have two linear ranges. Starting from low pressures (right-hand side), the first
linear range is fitted by (7.8), because the film is not very thick and the van der
Waals attraction of the molecules by the solid governs the adsorption process (FHH
regime). With rising pressure, at a critical film thickness of about $z_{crit} \approx 0.5$ nm
(equation (7.11)), the vapor–liquid surface tension γ becomes dominant and a step-
like increase of the number of particles adsorbed is observed. The fractal FHH theory
covers fractal dimensions of $d_s \approx 2.3$ up to a length scale of $z \approx 1$ nm, independent
of grade number. At this length scale a geometrical cut-off appears and the surface
becomes rougher. In the final linear regime, corresponding to $z > 1$ nm, the surface
fractal dimension takes the value $d_s \approx 2.6$ (CC regime). This linear range has an
upper cut-off length of $z \approx 6$ nm.

Figure 7.13 shows that, unlike the graphitized blacks, untreated furnace blacks
have only one linear range with a fractal dimension of $d_s \approx 2.6$ (CC regime, (7.9)).
Obviously the van der Waals attraction can be neglected and the surface tension γ
controls the adsorption process on all length scales. This is due to the larger surface
fractal dimension d_s compared with that of graphitized furnace blacks which shifts

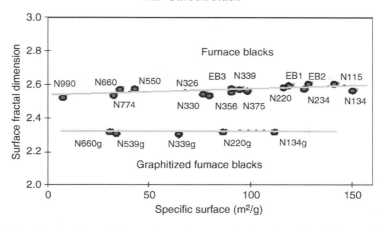

Fig. 7.14. Surface fractal dimensions d_s on atomic length scales of furnace blacks and graphitized blacks vs. the specific surface. The data are obtained from nitrogen adsorption isotherms in the multilayer regime. From [138].

the cross-over length z_{crit} to smaller values (equation (7.10)). Assuming that the number density n_p, the surface tension γ of the adsorbate, and the van der Waals interaction parameter α are approximately the same for liquid nitrogen adsorbed on graphitized and untreated furnace blacks, a cross-over length of $z_{crit} \approx 0.35$ nm can be estimated from (7.10) with the experimental values of the fractal dimensions and the cross-over length $z_{crit} \approx 0.5$ nm on a graphitized carbon black. The value $z_{crit} \approx 0.35$ nm is in the range of the detection limit given by the layer thickness $a \approx 0.35$ nm. Hence, the nitrogen adsorption on furnace carbon blacks is dominated by the vapor–liquid surface tension on all length scales and a cross-over between the FHH and the CC regime does not appear.

The results for the surface fractal dimension of a series of furnace blacks and graphitized blacks, obtained by nitrogen adsorption in the multilayer regime, are summarized in Fig. 7.14. The cut-off lengths are quite similar in both series of blacks and agree with those found in Figs. 7.12 and 7.13. In Fig. 7.14 a weak trend of increasing surface fractal dimension with increasing specific surface (decreasing primary particle size) is observed. This results from increasing curvature of the particle surface with decreasing size, since crystallite structures with edges are present on the surface that lead to a more pronounced roughness if arranged on a strongly curved surface. We will see in the next section that the number of crystallite edges and slit-shaped cavities increases slightly with increasing specific surface, leading to a higher energetic surface activity for the fine carbon blacks. This relatively small effect correlates well with the weak trend of the surface fractal dimension observed in Fig. 7.14.

The almost universal value of the surface fractal dimension $d_s \approx 2.6$ of furnace blacks can be traced back to the conditions of disordered surface growth during carbon black processing. It compares very well to the results evaluated within the anisotropic Kardar–Parisi–Zhang (KPZ) model [270] as well as numerical simulations of surface growth found for random deposition with surface relaxation. This is demonstrated in some detail in [166].

7.2.3 Energy distribution of carbon black surfaces

The energy distribution $f(Q)$ of carbon black surfaces is calculated by assuming that the measured overall isotherm consists of a sum of generalized Langmuir isotherms of various interaction energies Q, implying that the energy distribution can be identified with the numerically obtained weighting function [167, 168]. For a continuous distribution function $f(Q)$ the overall isotherm $\Theta(p, T)$ is given by

$$\Theta(p, T) = \int_0^\infty \theta(p, T, Q) f(Q) \, dQ \tag{7.12}$$

The integral in (7.12) is normalized to unity. This has to be taken into consideration if solid samples with different specific surface areas are compared.

For an evaluation of the local model isotherm $\theta(p, T, Q)$ with constant interaction energy Q, the effects of multilayer adsorption and lateral interactions between neighboring adsorbed molecules are considered by applying two modifications to the Langmuir isotherm: (i) a multilayer correction according to the well-known BET concept and (ii) a correction due to lateral interactions with neighboring gas molecules introduced by Fowler and Guggenheim (FG) [169]:

$$\theta(p, T, Q) = \frac{b_{\mathrm{BET}}^2 b_{\mathrm{FG}} b_{\mathrm{L}} p}{1 + b_{\mathrm{BET}} b_{\mathrm{FG}} b_{\mathrm{L}} p} \tag{7.13}$$

with

$$b_{\mathrm{BET}} = \frac{1}{1 - p/p_0}, \tag{7.14}$$

$$b_{\mathrm{FG}} = \exp\left(\frac{z\omega\theta}{RT}\right), \tag{7.15}$$

$$b_{\mathrm{L}} = \frac{N_A \sigma \tau_0}{\sqrt{2\pi MRT}} \exp\left(\frac{Q}{RT}\right). \tag{7.16}$$

Here, z is the number of neighboring adsorption sites, ω is the contribution of the lateral interaction to Q, θ is the probability that the neighboring sites are occupied by a gas molecule, R is the gas constant, T is temperature, τ_0 is the Frenkel's characteristic adsorption time, σ is the adsorption cross-section of the gas molecules,

N_A is Avogado's constant and M is the molar mass. Note that the probability that the neighboring sites are occupied by other gas molecules is taken to be the local surface coverage θ and not the overall surface coverage Θ. This means that sites with the same interaction energies Q are assumed to be arranged in patches, which is in accordance with the picture of graphite-like microcrystallites on the surface of the carbon black particles [3]. A probability Θ stands for a random distribution of sites with Q.

By referring to adsorption isotherms of ethene down to very low surface coverings (10^{-3}–1 monolayers), the energy distribution function of adsorption sites on different furnace blacks has been estimated with (7.12)–(7.16) by applying a numerical iteration procedure introduced by Adamson and Ling [167]. This is described in detail in [163]. For a test of the evaluation procedure, the resulting energy distribution functions obtained from four different isotherms (three different temperatures) of ethene on N220 are compared in Fig. 7.15. It is obvious that the isotherms measured at different temperatures lead to approximately the same result for the energy distribution function, confirming the applied procedure. An analysis of the energy distribution function of ethene on N220 is shown in Fig. 7.16, where the distribution function is fitted to four different Gauss functions.

Obviously, the good fit indicates that four different types of adsorption sites can be distinguished on the N220 surface. We relate the low-energetic peak I with $Q \approx 16$ kJ/mol to the basaltic layers and peak III with $Q \approx 25$ kJ/mol to the edges of carbon crystallites. Peak II with $Q \approx 20$ kJ/mol is related to amorphous carbon and peak IV with $Q \approx 30$ kJ/mol results from a few highly energetic slit-like cavities between carbon crystallites. This is shown schematically in Fig. 7.17.

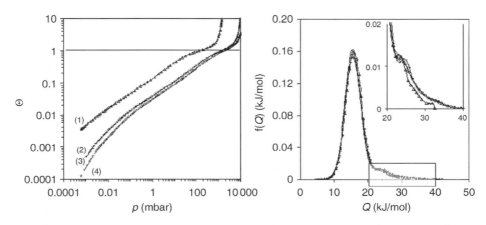

Fig. 7.15. Adsorption isotherms ($\Theta = N/N_m$) and corresponding energy distribution functions of ethene on N220 at various temperatures; ((1) $T = 177$ K; (2)(3) $T = 223$ K; (4) $T = 233$ K). From [138].

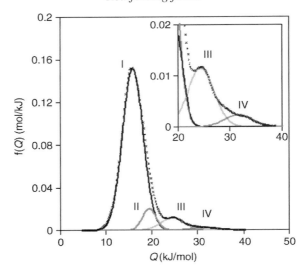

Fig. 7.16. Fitting of the energy distribution function of ethene on N220, already shown in Fig. 7.15 ($T = 223$ K), to four Gaussian peaks (I–IV). From [138].

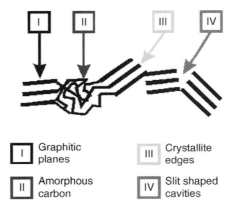

Fig. 7.17. Schematic view of the association between morphological arrangements of carbon crystallites and energetic characteristics of carbon black surfaces. Four different types of adsorption sites are distinguished that refer to the deconvolution shown in Fig. 7.16. From [138].

The attachment of peak II to the amorphous carbon is concluded from the observation that this peak does not appear with graphitized black N220g and graphitic powder. The corresponding isotherms and energy distribution functions are depicted in Fig. 7.18. A comparison with the above analyzed N220 and a strongly reinforcing channel gas black demonstrates the relatively large amount of highly energetic sites of these blacks (Fig. 7.18).

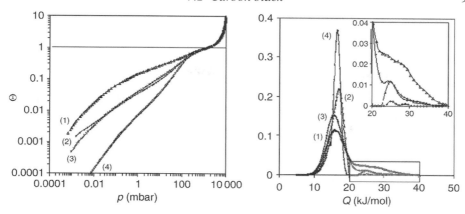

Fig. 7.18. Adsorption isotherms and evaluated energy distribution functions of ethene on four different colloidal fillers at $T = 223$ K; (1) channel gas black; (2) graphitic powder; (3) N220; (4) graphitized N220g. From [138].

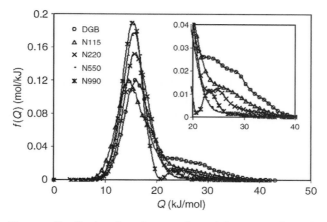

Fig. 7.19. Energy distribution functions evaluated from gas adsorption measurements of ethene at $T = 223$ K on different carbon black grades varying in particle size. From [163].

The energy distribution functions of different carbon blacks of varying particle size are shown in Fig. 7.19. Corresponding to the difference in level and shape of the isotherms the number of highly energetic sites varies significantly. The very fine Degussa gas black (DGB) and N115 have a large fraction of highly energetic sites, while N550 and N990, like the graphitized black N220g, show almost no highly energetic sites. The results of the peak analysis for the examined blacks and the graphitic powder are quantified in Table 7.1. More details about the energetic surface heterogeneity of carbon blacks can be found in [152, 156, 163, 170].

The analysis of the surface energy distribution has demonstrated that four different energetic sites can be distinguished on carbon black surfaces. The fraction

Table 7.1. *Estimated fraction (%) of the four different types of energetic sites (I–IV) for adsorption of ethene on various carbon blacks and graphite*

	I	II	III	IV
DGB	71	5	8	16
Channel gas black	61	2	31	6
N115	69	13	15	3
N220	84	7	7	2
N550	93	6	1	< 1
N990	96	0	3	1
Graphitic powder	94	0	4	2
Graphitized N220g	99	0	< 1	< 1

of highly energetic sites decreases significantly with grade number and disappears almost completely during graphitization. This indicates that the reinforcing potential of carbon black is closely related to the number of highly energetic sites, which can be well quantified by the applied gas adsorption technique. Theoretical investigations on the effect of morphological as well as energetic surface roughness on the polymer–filler interaction strength confirm this finding [44, 45, 171, 172]. Accordingly, the combination of two types of disorder, given by the pronounced morphological roughness ($d_s \approx 2.6$) and the inhomogeneous energetic surface structure of carbon blacks, enhances the polymer–filler coupling significantly. It represents an important reinforcing mechanism on atomic length scales associated with the required strong phase binding in high performance elastomer composites.

7.3 Silica

The basic characteristics of silica fillers, which can be altered during the precipitation process, are particle size distribution, porosity, specific surface area, and purity. Because of its high specific component of surface energy, silica has a stronger tendency to agglomerate than carbon black. The interactions between the polar groups (siloxane, silanol) on the surface of silica aggregates with non-polar groups of hydrocarbon elastomers are weak compared with the hydrogen-bonding interactions between surface silanol groups in silica itself. The dispersive forces between a non-polar rubber molecule and silica are low. For this reason, in principle, a hydrophobic modification of the silica surface would be expected to improve the compatibility of hydrocarbon elastomers and precipitated silica. Remarkable improvements in mechanical properties of silica-filled rubbers are obtained with the use of a coupling agent. The most widely used coupling agent today is bis-(triethoxysilylpropyl) tetrasulfide (TESPT). Extensive research has been done into several aspects of this agent, especially, into the steps involved in the reaction between silica and TESPT, as reported in [173–176]. TESPT is the most commonly used silane enabling silica to be used in tire compounds – tread compounds in particular. The use of TESPT is also the key factor in silica being a successful replacement for carbon black as the main reinforcing filler in the tread compound of the "green tire" [126].

If silica outperforms carbon black due to surface modification, the question arises of whether surface modification of carbon black can be used to enhance its performance. However, it appears that the improvement in the dynamic mechanical properties of carbon-black filled rubber by coupling agents is not as great as in the case of silica-filled vulcanizates [28]. Several factors are responsible for this difference. Unlike the silica surface, which is characterized by a uniform layer of siloxane and various types of silanol groups (isolated, geminal, and vicinal), carbon black surfaces contain not only hydrogen but also a number of different oxygen-containing groups (phenol, carboxyl, quinone, lactone, ketone, lactol, and pyrone). It is reasonable to expect that for a given coupling agent the different chemical functionalities would have different reactivities and, in effect, this has been found to be the case. Furthermore, the differences in the effectiveness of the coupling modification with carbon black and silica may also be associated with their microstructure. In the case of amorphous silica, the functional groups are randomly located on the surface, while the aggregates of carbon black consist of quasigraphitic crystallites, and the functional groups are located only at the edges of the graphitic basal planes of the crystallites. This suggests that as well as the difference in reactivity and concentration of the functional groups, the functional group distribution on the two filler surfaces is also different [28].

To understand better what happens during the early stage of mixing with silica, the role of silane in silica dispersion has to be studied, independently of the vulcanization chemistry and rubber–filler interactions. In [177] model systems of monodisperse silica aggregates, silanes, and squalene were studied. Small-angle neutron scattering (SANS) techniques were used to investigate the short-range arrangement of silica aggregates and the structure factor of the concentrated silica–squalene system. Using this technique it is possible to differentiate between the degrees of aggregation of particles through the interparticle distance.

At very low values of the wave vector q, which is related to large-scale interaction, a difference is noticed between the silica treated with TESPT, and silica treated with either octylsilane C18Et ($C_{18}H_{33}$–Si(OEt)$_3$) or TESPD (S_2–[C_3H_6–Si(OEt)$_3$] triethoxysilylpropyldisulfane). The difference suggests that the association of high agglomerates is stronger with TESPT than with TESPD. To confirm this observation, slurries of silane-treated silica were analyzed by light scattering. A significant difference in particle size could be observed between disulfide silane TESPD and

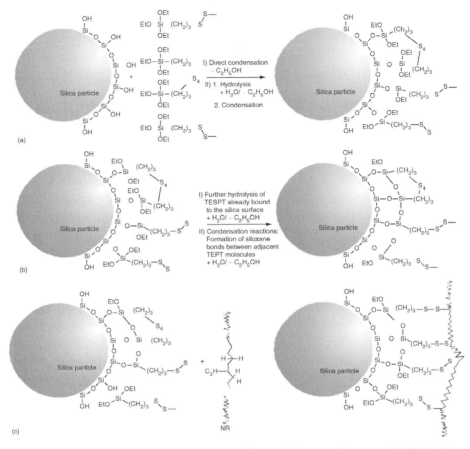

Fig. 7.20. (a) Primary reaction between silica and the coupling agent TESPT. (b) The two-step process of the secondary reaction between silica with coupling agent TESPT. (c) Possible reaction between TESPT and rubber.

tetrasulfide silane TESPT. The disulfide shows a lower diffraction at low angles, characteristic of smaller aggregates [177].

Coupling agents such as TESPT have triethoxysilyl functions which can react with the silanol groups on the silica surface. This so-called "silanization reaction" can be divided into primary and secondary reactions (Fig. 7.20). During the primary reaction, one ethoxy group of each Si-unit reacts with an accessible silanol group on the silica surface and, therefore, links chemically to the filler. Because of agglomeration of the silicas by vicinal hydrogen bonding, only a relatively small number of free OH groups is available. These agglomerates are partly dispersible under the high shear stress applied by the rotor of the mixer. This primary reaction may be followed, after a hydrolysis reaction, by a condensation reaction between

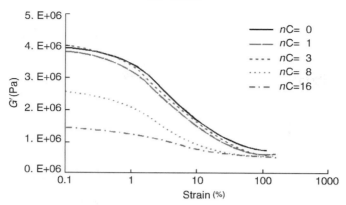

Fig. 7.21. Influence of the silane length on the Payne effect.

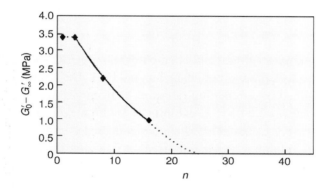

Fig. 7.22. Amplitude of the Payne effect as a function of silane length.

pairs of neighboring silane molecules already bound to the silica surface, which constitutes the secondary reaction. For mechanical reasons, the secondary reaction is always a two-step process: first the hydrolysis of one or two ethoxy groups (this reaction generates ethanol as a reaction product as well as one or two silanol groups, -SiOH), followed by a condensation reaction, which, in turn, generates either water or ethanol, depending on the state of hydrolysis of the silane atom reacting with the original silanol group. The water required for the intermediate hydrolysis step is adsorbed onto the silica surface.

We note that the properties of the final compound depend strongly on the length of the silane used and on the amount of coupling agent. Figure 7.21 shows the influence of the silane's length on the Payne effect, which describes the shear modulus G' as a function of strain amplitude for an S-SBR filled with 40 phr silica [178]. With increasing spacer length (the number of CH_2 units in the aliphatic chain bearing the siloxy groups), the large strain modulus G'_∞ remains roughly constant whereas

Reinforcing fillers

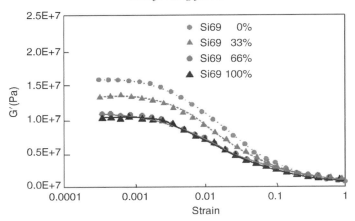

Fig. 7.23.   The influence of the surface coverage by silane (Si69) on the shear modulus as a function of the strain amplitude.

the small strain modulus G_0' gradually decreases. Figure 7.22 shows the amplitude of the Payne effect as a function of silane length [178]. The amplitude of variation of G' was determined by subtracting $G'(\gamma_0 = 1)$ from $G'(\gamma_0 = 10^{-4})$. The three shortest alkyl silanes ($n = 0, 1, 3$) exhibited very high $\Delta G'$. Further increase in spacer length ($n = 8, \ldots, 16$) induced a gradual decrease in the Payne effect. The occurrence of a minimum silane length necessary to alter the dynamic properties of the composites defined an apparent "length threshold" separating, as far as $\Delta G'$ is concerned, effective and ineffective molecules. Extrapolation of the data points revealed that the Payne effect may eventually vanish when n is increased to about 30. Figure 7.23 shows the influence of the amount of coupling agent on the shear modules as a function of strain amplitude [178].

As already discussed, the first step involved in the reaction between silica and TESPT has the effect of modifying the silica surface in such a manner that it becomes more hydrophobic, disrupting the silica–silica interactions, which, in turn, results in a partial disruption of the filler network. This is observed experimentally, for example, as a reduction in compound viscosity, which is a great benefit when producing these compounds in industrial conditions. The second major effect when using TESPT is the creation of permanent chemical bonds between modified silica particles and the polymer matrix (Fig. 7.20).

8

Hydrodynamic reinforcement of elastomers

8.1 Reminder: Einstein–Smallwood

In the following sections we are going to study the reinforcement obtained by adding particles to the elastic matrix. The mechanisms of the effective enhancement of the elastic modulus cannot be explained by one simple theory, since several interactions and many different length scales are involved [179]. This is because there are different physical levels of reinforcement. The rubber matrix contributes through its rubber elasticity [7], whereas the filler particles contribute in different ways. The most well known of these are volume effects, also called hydrodynamics interactions (due to the analogy with the enhancement of the viscosity of liquids by the addition of particles).

In the context of carbon-black-filled elastomers, the contribution to reinforcement on small scales can be attributed to the complex structure of the branched filler aggregates as well as to a strong surface–polymer interaction, leading to the so-called bound rubber. Thus the filler particles are coated with polymer chains and the binding (physical or chemical) of elastomer chains to the surface of the filler particles changes the elastic properties of the macroscopic material significantly [2]. On larger scales the hydrodynamic aspect of the reinforcement dominates the physical picture. Hydrodynamic reinforcement of elastic systems plays a major role not only in carbon-black-filled elastomers, but also in composite systems with hard and soft inclusions. Finally, at macroscopic length scales filler networking at medium and high filler volume fractions plays a dominant role [179].

In this chapter we are going to concentrate – on a general basis – on the different mechanisms of elastomer reinforcement in the hydrodynamic regime. To do so, we present two different regimes of reinforcement mechanisms. In order to introduce the subject, we briefly review the classical ideas. The simplest approach was presented as early as 1944 by Smallwood [180], who showed that adding randomly dispersed spherical filler particles to a rubbery matrix yields an elastic reinforcement of the form

$$G = G_m(1 + 2.5\phi). \tag{8.1}$$

This is known as the Einstein–Smallwood formula, where ϕ is the volume fraction of the filler components and G_m denotes the elastic modulus of the rubber matrix: these were calculated and discussed in Chapter 5. The physical conditions required for this result are: (i) freely dispersed particles, i. e. low volume fraction, (ii) a spherical shape (leading to the constant 2.5), and (iii) entirely non-elastic filler particles, i. e. their elastic modulus has to be infinitely large. Although the assumptions are very strict and idealized the Einstein–Smallwood equation contains essential physics. The reinforcement term contains two factors: one is a simple number which is related only to the geometry of the particles, the other is linear in the volume fraction of the filler particles. This latter point corresponds to a more general physical principle, as we will see later: as long as the filler particles do not overlap, this term will stay linear in this expansion.

In real systems none of these assumptions is rigorously valid, and many authors have provided generalizations of (8.1) by relaxing one or more of conditions (i)–(iii). For example particle–particle interaction at a less low filler volume fraction can be taken into account by an additional term quadratic in ϕ, which was first done by Guth and Gold [46]. However, most of these extensions have been carried out on a highly empirical basis. Here we want to show that interesting generalizations can also be obtained by rigorous theoretical calculations, thereby illuminating the main mechanisms of hydrodynamic reinforcement in complex composite systems. We are going to present two other extremes cases and shed light on the principal issues of the Einstein–Smallwood theory. First we are going to generalize the dispersion of the filler particles. As mentioned above, the linear dependence in (8.1) stems basically from the assumption of free dispersion of the filler particles. Only in this case are the conditions for an expansion in terms of the volume fraction given. If the filler particles form clusters, two different regimes must be distinguished. First, small clusters may be dispersed freely without contact between them. Then we can expect a reinforcement law as given by Einstein–Smallwood to hold. The only changes to be expected are in the geometry factor, which may become size-dependent. The second regime comes into play when the clusters begin to overlap. Then a different behavior as a function of the volume fraction ϕ is to be expected. If it is assumed that the clusters may have some "fractal" geometry, then some scaling relations can be derived; see Section 8.2. In the overlap regime a stronger reinforcement is expected.

The other extreme generalization of the Einstein–Smallwood law again assumes free dispersion, but with the filler particles having more complex elastic properties.

In Section 8.3 we will present a detailed theory on so-called core–shell systems. To do so, we again assume the shape of the filler particles to be spherical, but now they may have, e.g., a hard core of a different elastic modulus and a softer shell. The second assumption in Section 8.3 is that these core–shell filler particles are distributed randomly in the elastic matrix, i.e. we assume they do not cluster. Alternatively the theory can be kept general, so that filler particles with holes ("swiss-cheese"-like solids) can also be discussed.

8.2 Rigid filler aggregates with fractal structure

Deviations from the Einstein–Smallwood formula for non-spherical filler particles, e.g., carbon black aggregates, so far have mostly been discussed in terms of an effective volume fraction $\phi_{\text{eff}} > \phi$ [181, 182] while retaining the form of (8.1). Carbon black aggregates are built up from approximately spherical primary particles which are connected in a branched but nevertheless solid way, see Fig. 8.1. These structures are known to have universal features and thus can be characterized by the fractal exponents d_{f} (the mass fractal dimension) and D (the spectral dimension as a measure of aggregate connectivity) [181, 183].

The branched aggregate structure leads to the possibility of aggregate overlap in the case of medium and high filler volume fractions. We expect this to affect not only the factor ϕ_{eff}, but also the scaling behavior of the ϕ dependence. Therefore we use an elastic theory to calculate – independently of the Einstein–Smallwood formula – the dependence of the modulus on the universal fractal structure of carbon black

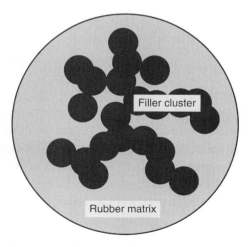

Fig. 8.1. Schematic view of the carbon black aggregate structure, which can be characterized by fractal exponents.

aggregates [179]. The calculations are, however, non-trivial and require methods from effective medium theories [184, 185]. The mathematical formalism requires some ideas from the general theory of Green's functions. Indeed, the formalism presented in [184, 185] can be extended to more general shapes of fractal filler agglomerates (or filler clusters of a certain shape which satisfy on reasonable scales certain scaling relations regarding their size and their connectivity). The filler particles which have a certain shape interact with the elastic matrix and eventually at concentrations larger than the overlap concentration with themselves. For the small deformations considered here, we can assume a perfect binding of the matrix to the filler surfaces. Furthermore, the interaction between filler aggregates is not taken into account explicitly, the matrix is considered as incompressible and ideally elastic. The many-particle effects which occur at higher concentrations are dealt with in the framework of an effective medium theory, the so-called "self-consistent screening approximation" [185, 186].

To pursue this idea we have first to generalize the mathematical formulation of the statistics of linear polymers from Chapter 2 (see (2.16)) to general fractal objects. We may assume that the clusters formed by the filler particles can be described by a fractal shape. We do so only to describe the structure, rather than suggesting that the structure is fractal. It will be obvious shortly that this assumption allows us to predict certain specific forms of the reinforcement. First, it allows us to introduce an effective probability distribution for the filler clusters [187, 188]:

$$P[\mathbf{R}(\mathbf{s})] \propto \exp\left\{ -\frac{3}{2b^2} \int_0^L \mathrm{d}^D s \left(\frac{\partial \mathbf{R}(\mathbf{s})}{\partial \mathbf{s}} \right)^2 \right\} \tag{8.2}$$

$$\propto \exp\left\{ -\frac{3}{4\pi b^2} \sum_{\mathbf{p}} |\mathbf{p}|^2 |\mathbf{R}_{\mathbf{p}}|^2 \right\}, \tag{8.3}$$

where b is the size of the individual filler particles (primary particle). The spectral dimension D describes the connectivity of the filler particles, e.g., linearly connected objects such as polymer chains or random walks correspond to $D = 1$, percolation clusters correspond roughly to $D = 4/3$. Here $\mathbf{R}(s)$ describes any spatial vector which points to the object. For our purpose it is useful to work in Fourier space. Therefore we introduce the transform $\mathbf{R}(s) = \sum_{\mathbf{p}} \mathbf{R}_{\mathbf{p}} \exp(i s \cdot \mathbf{p})$ by analytic continuation for arbitrary spectral dimensions. Thus $\mathbf{p}$ is the Fourier conjugate of the internal space variable $\mathbf{s}$.

However, we need modeling of the filler structure, which requires a further generalization to self-avoidance. In this case we may generalize the distribution

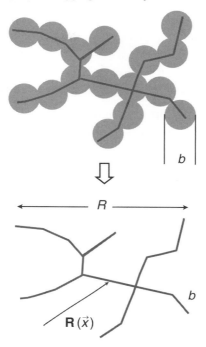

Fig. 8.2. Coarse graining of the filler aggregate structure leads to a mathematical description similar to the case of branched polymers with excluded volume: b denotes the mean primary particle diameter, R the mean aggregate size.

due to [188]:

$$P[\mathbf{R(s)}] \sim \exp\left\{-\frac{3}{4\pi b^2}\sum_{\mathbf{p}}|\mathbf{p}|^{2\alpha}|\mathbf{R_p}|^2\right\},\qquad(8.4)$$

where $\alpha = D(2+d_f)/2d_f$ and b is the effective diameter of primary particles. This Gaussian distribution results from a coarse graining which produces a structure similar to the case of branched polymers with excluded volume, see Fig. 8.2. Note that for the choice of $D = 1$ and $d_f = 2$ the above equation reduces to the probability distribution of random-walk-shaped curves.

The effective exponent $\alpha \neq 1$ can be expressed in terms of the corresponding fractal dimensions, i.e. $\langle\mathbf{R}^2\rangle \sim L^{2\alpha-D} \overset{!}{\sim} L^{2\nu}$ or $d_f = 2D/(2\alpha - D)$ which yields $\alpha = \nu + D/2$ or alternatively with the excluded volume-exponent $\nu = D/d_f = 2/d_w$, where d_w is the dimension that defines the distance that a random walker travels on the fractal $R^{d_w} \propto t$.

Furthermore, we note the more general relations (see [188] for a review) $\nu = (2 - D)/2$ for ideal phantom clusters (where the branches can penetrate) and, more realistically, $\nu = (D + 2)/(d + 2)$ for clusters with excluded volume. The next step is to calculate the self-energy function (see Section 8.2.1), which corresponds directly to the reinforcement factor.

8.2.1 Effective medium theory and linear elasticity

In this subsection we derive the equations for a linear isotropic elastic material with an elastic modulus μ, compression modulus B, Poisson number $\sigma = (3B - 2\mu)/(6B + 2\mu)$, and Young's modulus $G_m = 2\mu(1 + \sigma)$. Since the material is assumed to be almost incompressible, i.e. B is very large [6], we may assume that the rubber matrix is effectively incompressible for the purpose of the present chapter, i.e. we assume a very large compression modulus, $B \to \infty$, which defines the Poisson ratio $\sigma = \frac{1}{2}$ for such incompressible materials. Furthermore, we assume there are only small deformations and we can therefore compute only the initial modulus, i.e. we ignore finite extensibility effects, as in earlier studies, e.g. [7], the results of which have been confirmed many times [96, 189–191]. The other advantage of a theory for small deformations is that we do not have to worry about the rupture of filler particles from the elastic matrix. These assumptions allow us now to extend the theory used by Cates and Edwards [185] and Freed and Edwards [186].

We begin with a simple illustration of the theory. The local deformation field of a filled system can be described as

$$\mathbf{u}(\mathbf{r}) = \int d^3r' \, \mathbf{G}_0(\mathbf{r} - \mathbf{r}') \cdot \mathbf{F}(\mathbf{r}') + \int ds \int d^3r' \, \delta(\mathbf{R}(s) - \mathbf{r}')\mathbf{G}_0(\mathbf{r} - \mathbf{r}') \cdot \boldsymbol{\sigma}(s),$$

(8.5)

where the Green elastic tensor (the Green function of the unfilled matrix) is given by [192]

$$\mathbf{G}_0(\mathbf{r}) = \frac{1}{16\pi\mu|\mathbf{r}|} \left\{ \left(\frac{3 - 4\sigma}{1 - \sigma}\right)\mathbf{I} + \frac{\hat{\mathbf{r}}\hat{\mathbf{r}}}{1 - \sigma} \right\}$$

(8.6)

and $\boldsymbol{\sigma}(s)$ describes a local stress field. The general shape of the filler particles is defined by the parameterization $\mathbf{R}(s)$ (for simplicity assumed to be one-dimensional connectivity here, $D = 1$), and $\mathbf{F}(\mathbf{r})$ is the (small) elastic force. The symbol $\mathbf{I}$ denotes the unit matrix and $\hat{\mathbf{r}}\hat{\mathbf{r}}$ is the matrix defined by the unit vectors $\hat{\mathbf{r}}$ which acts as projection on the direction of $\mathbf{r}$, see [184, 192] for details. The essential problem is now to calculate a new form of the Green function $\mathbf{G}$ which contains the effects of the filler particles. Therefore all effects have to be taken into account: the shape of the filler particles, the spatial distribution of the particles, etc. Detailed formalisms for these calculations have been developed in effective medium theories and we can just summarize here.

It turns out that it is useful to work in Fourier space. The desired quantity is then given by the average local deformation:

$$\langle \mathbf{u}(\mathbf{k}) \rangle = \mathbf{G}(\mathbf{k}) \cdot \mathbf{F}(\mathbf{k}) , \tag{8.7}$$

where the Green function of the filled system is given by

$$\mathbf{G}(\mathbf{k}) = \mathbf{G}_0(\mathbf{k}) - \mathbf{G}_0(\mathbf{k}) \left\langle \sum_q \phi^2(\mathbf{k}, q) \mathbf{\Delta}(q) \mathcal{G}_0(q) \right\rangle \mathbf{G}_0(\mathbf{k}) , \tag{8.8}$$

in which the argument q corresponds to Fourier components describing the shape of the particles, i. e. the transformation of $\mathbf{R}(s)$ to the corresponding representation in Fourier components $\mathbf{R}(q)$. The symbol $\mathbf{\Delta}(q) = \Delta G_q^2 \hat{\mathbf{n}}\hat{\mathbf{n}}$ is the Fourier transform of the deformation operator and $\mathcal{G}_0(q)$ is the general Green function of the shape of the filler particle. ΔG is the difference between the moduli of the filler particle and the matrix. Thus we have

$$\sum_{q'} (\delta_{qq'} \mathbf{I} + [q|\mathbf{G}_0|q'] \mathbf{\Delta}(q')) \mathcal{G}_0(q) = \mathbf{I} . \tag{8.9}$$

The reinforcement effects can then be written on a general basis as

$$\mathbf{G}(\mathbf{k}) = \mathbf{G}_0(\mathbf{k}) - \mathbf{G}_0(\mathbf{k}) \langle \mathbf{T} \rangle \, \mathbf{G}_0(\mathbf{k}), \tag{8.10}$$

where $\langle \mathbf{T} \rangle$ contains the information on the filler particles and can be written as

$$\mathbf{T} = \phi \mathbf{\Delta} \mathcal{G}_0 \phi = \sum_q \phi^2(\mathbf{k}, q) \, \mathbf{\Delta}(q) \, \mathcal{G}_0(q). \tag{8.11}$$

The quantity ϕ contains all the information on the shape (and elasticity) of the filler particles.

These approaches can be generalized to many filler particles. The natural generalization of (8.10) is

$$\mathbf{G} = \mathbf{G}_0 - \mathbf{G}_0 \sum_{\alpha=1}^{N} \langle \mathbf{T}^\alpha \rangle \mathbf{G}_0 + \mathbf{G}_0 \sum_{\alpha \neq \beta} \langle \mathbf{T}^\alpha \mathbf{G}_0 \mathbf{T}^\beta \rangle \mathbf{G}_0$$

$$- \mathbf{G}_0 \sum_{\alpha \neq \beta} \sum_{\gamma \neq \beta} \langle \mathbf{T}^\alpha \mathbf{G}_0 \mathbf{T}^\beta \mathbf{G}_0 \mathbf{T}^\gamma \rangle \mathbf{G}_0 + \cdots , \tag{8.12}$$

where α, β, γ are the numbers of filler particles in the matrix. Note that (8.12) is of the Dyson type, which is well known in quantum field theory. Indeed, the methods developed there are useful to resolve the present problem at least in some reasonable approximations. It is most convenient to recast it in a more comfortable form in terms of the geometric series:

$$\mathbf{G}^{-1} \approx \left(\mathbf{G}_0 - \mathbf{G}_0\bar{\mathbf{T}}\mathbf{G}_0 + \mathbf{G}_0\bar{\mathbf{T}}\mathbf{G}_0\bar{\mathbf{T}}\mathbf{G}_0 - \cdots\right)^{-1} = \mathbf{G}_0^{-1} + \bar{\mathbf{T}}. \tag{8.13}$$

As long as the filler particles do not overlap and do not interact with each other the solution of (8.12) yields a simple generalization of the Einstein–Smallwood law. If the particles overlap, interactions may screen themselves, as is well known from the theory of concentrated polymer solutions [62]. Indeed, it turns out here that similar methods such as a "self-consistent screening approximation" can be used to solve the problem of many filler particles above their overlap concentration. This is obvious if the filler particles can interpenetrate each other, such as in fractal-like aggregates. Then the reinforcement is not only given by the volume effect, but also by the overlap-induced additional reinforcement. The theory of Green functions allows us to compute the corresponding self-energy Σ in terms of the irreducible diagrams in the matrix $\bar{\mathbf{T}}$. The self-consistent screening model leads to a set of equations which have to be solved self-consistently and simultaneously:

$$\mathbf{G}^{-1}(\mathbf{k}) = \mathbf{G}_0^{-1}(\mathbf{k}) + \sigma(\mathbf{k}) , \tag{8.14}$$

$$\Sigma(\mathbf{k}) = N\left\langle \sum_q \phi^2(\mathbf{k},q)\Delta(q)\mathcal{G}(q) \right\rangle , \tag{8.15}$$

$$\mathbf{I} = \sum_{q'}\left(\delta_{qq'}\mathbf{I} + [q|\mathbf{G}|q']\Delta(q')\right)\mathcal{G}(q) , \tag{8.16}$$

where $\mathbf{1}$ is the unit matrix. The most interesting effect is the mode dependence of the effective screening, which shows that the screening of the elastic interaction is only relevant for small length scales, while on large scales the elasticity becomes reinforced. We may then approximate the screened elastic Green function by

$$G(\mathbf{k}) \propto \frac{1}{\mu(k^2 + \xi^{-2})} , \tag{8.17}$$

where we have denoted the screening length by ξ. In the following we need to compute the screening lengths as a function of filler shape and concentration. The next step is then to evaluate the corresponding properties for filler agglomerates, i. e. to compute these functions with respect to the properties of the filler particle.

First, we note that the shape function has the general form

$$\left\langle \phi^2(\mathbf{k},q) \right\rangle \propto \frac{(kb)^{(2-D)/\nu}}{(k^2b^2\,C_1)^{1/\nu} + q^2}, \tag{8.18}$$

which is easily calculated using the distribution equation (8.4) and the decomposition into "Rouse modes." Then the self-energy can be written as

$$\Sigma(\mathbf{k}) \propto c\,(kb)^{(2-D)/\nu} \sum_{\mathbf{q}} \frac{\{K(q)\}^{-1}}{(k^2 b^2\,C_1)^{1/\nu} + q^2}, \tag{8.19}$$

where c is the filler concentration and $K(\mathbf{q})^{-1} = \mathbf{\Delta}(\mathbf{q}) \cdot \mathbf{G}(\mathbf{q})$. Equation (8.19) can be evaluated further and yields different results for the overlap and non-overlap regimes. To see this, the evaluation of the self-energy has to be analyzed for the overlap criterion above the overlap concentration $c \sim N/R^3$, where N is the number of primary particles of size b in the cluster and R its average extension. The clusters overlap only if their connectivity is not too large. Criteria for these conditions are worked out in [116, 193]. It was shown that the spectral dimensions should not exceed 6/5 for ideal clusters and 4/3 for non-ideal clusters. Clusters with larger connectivities do no longer overlap significantly. Then the self-energy can be written as

$$\Sigma(\mathbf{k}) \overset{k\to 0}{\propto} \begin{cases} c\,\mu\,b^{-2}\,(kb)^{(2-D)/\nu}\,L^{2+\nu-2D} & \text{without overlap } \xi \geq bL^{\nu} \\ c\,\mu\,b^{-2}\,(kb)^{(2-D)/\nu}\,L^{2-D}(\xi/b)^{1-D/\nu} & \text{with overlap } \xi \ll bL^{\nu} \end{cases}, \tag{8.20}$$

where L is the mean (linear) cluster diameter.

8.2.2 Screening lengths

The last step in our approach is the computation of the screening length, which will introduce the cluster concentration (or volume fraction). The screening length can be estimated by simple scaling arguments or alternatively by the use of the general theory that we outline in the following subsection. Both methods yield the same results and we restrict ourselves to presenting the scaling estimates.

To do so, we note that the overlap concentration for the clusters is given by

$$c^* = \frac{b^3 L^D}{R^3} = \frac{b^3 L^D}{b^3 L^{d\nu}} = L^{D-3\nu} \tag{8.21}$$

and assume that the screening length has (as in the theory of polymer solutions) the simple scaling form

$$\xi = R\,f\!\left(\frac{c}{c^*}\right) = R\left(\frac{c}{c^*}\right)^x = bL^{\nu}\left(\frac{c}{c^*}\right)^x. \tag{8.22}$$

Simple geometrical arguments and dimensional counting yield the result

$$
\xi = \begin{cases}
b\,c^{1/(d_f-3)} = b\,c^{v/(D-3v)} & \text{general} \\
b\,c^{-(2-D)/(6-5D)} & \text{for ideal clusters } D < 6/5 \\
b\,c^{-(D+2)/[2(3-D)]} & \text{for swollen clusters}
\end{cases}
\tag{8.23}
$$

The scaling results can be confirmed by calculating the self-energy and we simply note the result

$$
\Sigma(\mathbf{k}) \sim c\,\mu\,b^{-2}\,(\xi/b)^{1-D/v},
\tag{8.24}
$$

where ξ must be of the same form as predicted by the scaling arguments since we must have $\Sigma \sim \mu\xi^{-2}$.

8.2.3 Reinforcement by fractal aggregates

The results for the screening length yield the reinforcement factor as a function of the volume fraction $\phi = b^3 c$. To bring the results into a more useful form we replace the linear cluster size L by its spatial dimension $R = bL^v$ in (8.20). Then using this and $G_0(\mathbf{k}) \sim \mu^{-1}k^{-2}$ we obtain the general result:

$$
\frac{G - G_{\mathrm{m}}}{G_{\mathrm{m}}} \overset{k\to 0}{\sim} (kb)^{(2-2v-D)/v}
$$

$$
\times \begin{cases}
\left(\dfrac{R}{b}\right)^{(2+v-2D)/v} c & \text{below overlap concentration} \\[2mm]
\left(\dfrac{R}{b}\right)^{(2-D)/v} c^{\frac{2v}{3v-D}} & \text{above overlap concentration}
\end{cases}
\tag{8.25}
$$

or alternatively using $v = D/d_{\mathrm{f}}$

$$
\frac{G - G_{\mathrm{m}}}{G_{\mathrm{m}}} \overset{k\to 0}{\sim} (kb)^{2(d_{\mathrm{f}}/D)-d_{\mathrm{f}}-2}
$$

$$
\times \begin{cases}
\left(\dfrac{R}{b}\right)^{2(d_{\mathrm{f}}/D)-2d_{\mathrm{f}}+1} \phi & \text{no overlap } \phi \le \left(\dfrac{R}{b}\right)^{d_{\mathrm{f}}-3} \\[2mm]
\left(\dfrac{R}{b}\right)^{2(d_{\mathrm{f}}/D)-d_{\mathrm{f}}} \phi^{2/(3-d_{\mathrm{f}})} & \text{with overlap } \phi \gg \left(\dfrac{R}{b}\right)^{d_{\mathrm{f}}-3}
\end{cases}
\tag{8.26}
$$

For a realistic modeling of primary carbon black aggregates by ballistic cluster–cluster aggregation with $d_{\mathrm{f}} = 1.9$ and $D = 1.3$ [141–143] (see Section 7.2), we find

$$
\frac{G - G_{\mathrm{m}}}{G_{\mathrm{m}}} \sim \begin{cases}
R^{0.1}\,\phi & \text{for } \phi < \phi_{\mathrm{crit}} & \text{(a)} \\
R^{1.0}\,\phi^{1.8} & \text{for } \phi > \phi_{\mathrm{crit}} & \text{(b)}
\end{cases}
\tag{8.27}
$$

$\phi_{\text{crit}} = (R/b)^{1.1}$ denotes the critical overlap volume fraction for the branched filler aggregates. Thus the two different regimes correspond to (a) non-overlapping clusters and (b) overlapping clusters, depending on filler volume fraction. In the non-overlapping regime, the behavior is similar to (8.1) as we guessed earlier, i.e. the reinforcement is proportional to the volume fraction, whereas in regime (b) the hydrodynamic reinforcement sensitively depends on the universal aggregate structure.

In the non-overlapping regime (a), the filler contribution to the modulus is always proportional to the filler concentration itself and a geometrical factor. Due to the fractal nature of the filler aggregates, this factor depends on the mean aggregate size. This stems from the general concept of fractal elasticity [183]. From (8.27) we determine the aggregate size dependence of the reinforcement to be weak without overlap, but almost linear with overlap. This again is a result of the branched structure of the filler aggregates. The disadvantages of this model are the small range of application and the idealizations which we introduced in order to make the calculations tractable. The advantages are the successful derivation of a structure–property relationship, the possibility of explicitly including the fractal filler structure, and the universality (transfer to all types of branched aggregates). Refinements of the present model require the inclusion of local properties, such as particle–particle binding between the primary filler particles.

The dependence of the hydrodynamic reinforcement contribution on the universal aggregate structure is found to be weak at small filler concentrations, but strong for high filler content. Similar results have been found and confirmed by experimental data [123].

8.3 Core–shell systems

Now we return to spherical filler particles, but relax the condition of filler particle stiffness. Thus we assume that the particles are still freely dispersed, but themselves have elasticity. Examples of such filler particles are elastic microgels or lattices. The general theory for such systems has been derived by Felderhof and Iske and the theoretical details can be found in [194]. Their general result for the effective shear modulus is

$$\frac{G}{G_{\text{m}}} = 1 + \frac{[\mu]\,\phi}{1 - \frac{2}{5}[\mu]\phi}. \tag{8.28}$$

This results from a mean field approximation, which corresponds to the Lorentz local field method in the theory of dielectrics, leading to the famous Clausius–Mosotti equation for the effective dielectric constant. For rigid and spherical filler particles at low volume fraction, the Einstein–Smallwood formula is recovered because $[\mu] = 5/2$ (the intrinsic modulus $[\mu]$ follows from the solution of a single-particle problem). But the result clearly goes beyond the limits of Einstein–Smallwood, since two-body interaction (excluded volume) is included, leading to the strong increase of the modulus at high volume fraction.

Thus (8.28) provides a useful framework for the investigation of specific composite systems with spherically symmetric filler geometry: only the intrinsic modulus $[\mu]$ remains to be calculated. Along the lines of Jones and Schmitz [195], this can be done from the hydrostatic equilibrium equation for one particle included in a continuous elastic medium. In general the equilibrium equation can only be solved numerically. Exact analytical results are obtained in the following cases.

8.3.1 Uniform soft sphere

The simplest model consists of randomly dispersed uniform soft spheres. There are two limiting cases: if the modulus of the soft filler particles is zero, the matrix contains holes (resembling a Swiss cheese) and thus becomes softer. Such a material may seem to be only of theoretical interest, but it nevertheless will show how the theory works. On the other hand, in the case of a very large modulus of the filler particles, the Einstein–Smallwood formula will be reproduced. For uniform soft filler particles with elastic modulus $G_f > G_m$ there are several methods to calculate the intrinsic modulus, for a review see the book of Christensen [196]. The result as given by Jones and Schmitz [195] is

$$[\mu] = 5 \frac{1 - G_m/G_f}{2 + 3G_m/G_f}. \tag{8.29}$$

Inserting this into (8.28) leads to

$$\frac{G}{G_m} = 1 + \frac{5}{2} \phi \frac{G_f/G_m - 1}{G_f/G_m + 3/2 - \phi(G_f/G_m - 1)}, \tag{8.30}$$

plots of which are shown in Fig. 8.3.

This is identical with previous results (as reviewed by Christensen) in first order of ϕ and is valid also for intermediate volume fractions. As these results are already well known, further discussion will be omitted here.

8.3.2 Soft core/hard shell

We now investigate the case of a particle with a soft core (whose modulus is taken as zero for simplicity) and a shell with a modulus G_{shell} that is always larger than the modulus of the matrix, $G_{shell} \geq G_m$, see Fig. 8.4.

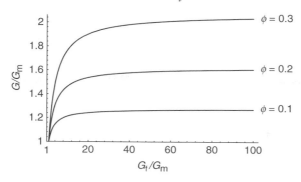

Fig. 8.3. Uniform soft filler particles with elastic modulus G_f: relative increase of the elastic modulus as a function of the ratio G_f/G_m for different values of the filler volume fraction. Reprinted from [45] with permission from the ACS.

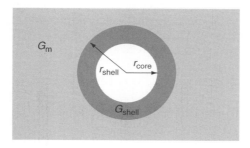

Fig. 8.4. Geometry and nomenclature for core–shell systems. G_m denotes the matrix modulus, G_{shell} the shell modulus and r_{shell} and r_{core} denote the sizes of the particles.

The algebraic expressions for the intrinsic modulus is

$$
\begin{aligned}
[\mu] = {}& -\frac{5}{3} + \frac{25}{3}\tilde{\mu}\left\{16\tilde{\mu} + 19 + 5(8\tilde{\mu} - 15)\tilde{r}^3 \right.\\
& + (\tilde{\mu} - 1)(-112\tilde{r}^5 + 75\tilde{r}^7 - 19\tilde{r}^{10})\Big\} / \Big\{(3\tilde{\mu} + 2)(16\tilde{\mu} + 19)\\
& + (\tilde{\mu} - 1) \times \left[50(4\tilde{\mu} + 3)\tilde{r}^3 - 112(3\tilde{\mu} + 2)\tilde{r}^5\right.\\
& \left.\left. + 75(3\tilde{\mu} + 2)\tilde{r}^7 + 38(\tilde{\mu} - 1)\tilde{r}^{10}\right]\right\},
\end{aligned}
\tag{8.31}
$$

where $\tilde{\mu} = G_m/G_{shell}$ and $\tilde{r}\tilde{r} = r_{core}/r_{shell} = r_{shell}/r_{core}$. The resulting effective modulus of the system is depicted in Figs. 8.5 and 8.6.

As expected, the result depends on the relation between the shell modulus and the matrix modulus G_{shell}/G_m and on the ratio of the outer and inner shell radii (as a measure of shell thickness) only. For finite G_{shell}/G_m, the whole system

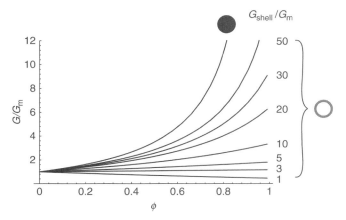

Fig. 8.5. Filler particles with a soft core: the relative increase of the elastic modulus as a function of filler volume fraction for different values of the ratio G_{shell}/G_m. The ratio of the outer to the inner shell radius is taken as $4/3$, the black circle denotes the limiting case of a totally rigid shell, i. e. $G_{shell}/G_m \to \infty$. Reprinted from [45] with permission from the ACS.

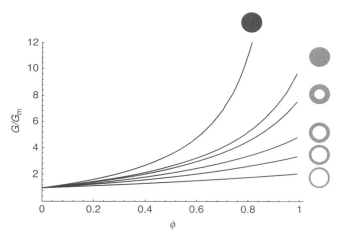

Fig. 8.6. Filler particles with a soft core: the relative increase of the elastic modulus as a function of filler volume fraction for different values of the ratio of the outer to the inner shell radius, from bottom to top $6/5$, $4/3$, $3/2$, 2, and ∞. The ratio G_{shell}/G_m is taken as 10, the black filled circle denotes the limiting case of a totally rigid shell, i. e. $G_{shell}/G_m \to \infty$. Reprinted from [45] with permission from the ACS.

remains elastic, i. e. there is no divergence of the effective elastic modulus for ϕ approaching 1. As can be seen from Fig. 8.5, reinforcement takes place only if the stiffness of the shell compensates for the softness of the core. This reinforcement condition is presented in a more general way in Fig. 8.7.

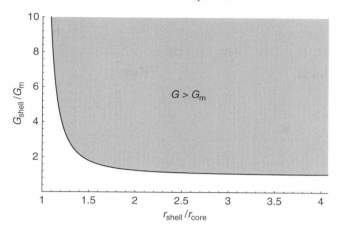

Fig. 8.7. Reinforcement condition for a soft-core–hard-shell system: reinforcement takes place only for parameters in the shaded region.

8.3.3 Hard core/soft shell

In the same way, the intrinsic modulus can be calculated for filler particles with a totally rigid core (with an infinitely large modulus) and a soft shell. If the shell modulus G_{shell} is assumed to be larger than the matrix modulus G_m, this system resembles a carbon-black-filled rubber, where the carbon black particles are surrounded by a bound rubber layer. Here the algebraic expression for the intrinsic modulus reads

$$
\begin{aligned}
[\mu] = \frac{5}{2} - 25\tilde{\mu} \times & \Big\{ (1 - \tilde{r}^3)(8\tilde{\mu} + 19/2) + (\tilde{\mu} - 1)(-42\tilde{r}^3 \\
& + 84\tilde{r}^5 - 50\tilde{r}^7 + 8\tilde{r}^{10}) \Big\} \Big\{ (3\tilde{\mu} + 2)(16\tilde{\mu} + 19) \\
& + (\tilde{\mu} - 1) \times \Big[-300(\tilde{\mu} + 3/4)\tilde{r}^3 + 168(3\tilde{\mu} + 2)\tilde{r}^5 \\
& - 100(3\tilde{\mu} + 2)\tilde{r}^7 + 48(\tilde{\mu} - 1)\tilde{r}^{10} \Big] \Big\},
\end{aligned}
\tag{8.32}
$$

where $\tilde{\mu} = G_m/G_{shell}$ and $\tilde{r}^{-1} = r_{shell}/r_{core}$. Figures 8.8 and 8.9 show the resulting effective modulus.

Most interesting is the large increase in reinforcement even for small bound rubber thicknesses (Fig. 8.9). In spite of the fractal filler structure as well as many-particle interactions being neglected, the curves have quite realistic features. Unfortunately, for comparison with experimental data values for the effective bound rubber thickness and strength are still lacking. To obtain and insert these seems worthwhile as the theoretical curves do not contain any fit parameters, i. e. all the parameters that result from structural filler and matrix properties.

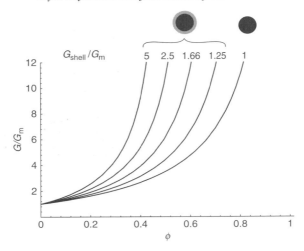

Fig. 8.8. Filler particles with a hard core: the relative increase of the elastic modulus as a function of filler volume fraction for different values of the ratio G_{shell}/G_m. The ratio of the shell radius to the core radius is taken as $4/3$. Reprinted from [45] with permission from the ACS.

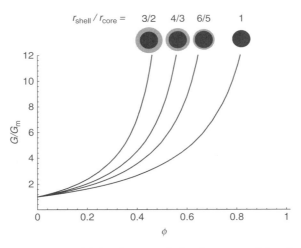

Fig. 8.9. Filler particles with hard core: relative increase of the elastic modulus as a function of filler volume fraction for different values of the ratio of the shell radius to the core radius. The ratio G_{shell}/G_m is taken as 2. Reprinted from [45] with permission from the ACS.

So far two cases have been studied in detail. The first is the reinforcement by rigid (fractal) aggregates and agglomerates of filler particles. The second is the reinforcement by randomly dispersed core–shell particles containing at least one soft component. The theoretical methods used are extensions of previously developed formalisms on continuum elasticity of composite materials. They differ in detail but are based on the same principles of elastic theory.

Let us briefly discuss the advantages of the model. The results obtained are realistic for small as well as intermediate filler concentrations, i. e. they are in accordance with experiments at least qualitatively. For the core–shell systems we have provided exact calculations of intrinsic moduli for various special forms of core–shell elasticity, i. e. soft spheres, hard spheres with soft surfaces etc. These results contain no fit parameters and in principle both compressible and incompressible media are accessible (only the incompressible case was shown here).

9

Polymer–filler interactions

9.1 General remarks and scaling

Although understanding the behavior of polymers on heterogeneous surfaces is a general problem in theoretical physics it provides deep insight into the problem of reinforcement and contributions. It is well accepted that the filler particles form large clusters which diffuse throughout the mixture to provide the most significant reinforcement effect on large macroscopic scales [179, 181, 197, 198]. Consequently these clusters form large surfaces inside the elastomer and allow significant polymer–filler contacts. Figure 9.1 shows a typical particle aggregate, with its hierarchy of length scales. The aggregate consists of individual particles, each with an irregular rough surface. As the particles form larger aggregates the irregular surfaces become very large. Moreover, the aggregates themselves form large clusters when the filler concentrations are high enough. Therefore we can expect major contributions to the reinforcement from the interaction between the polymer matrix and the irregular, rough surfaces.

However, the filler particles do not have homogeneous surfaces, but are strongly disordered. The disorder can be categorized in two extreme cases. In the first, the filler particles are spatially disordered. The second extreme case arises from the irregularity of the interactions. Imagine the surface to be spatially flat, but with the interaction energy varying randomly at each point on the surface. Such surfaces show non-trivial effects on the surrounding polymers as well. Both cases are driven by typical "disorder effects," which we will study in Section 9.2. Indeed several studies [135, 156, 199] suggest a strongly heterogeneous surface. Gerspacher and coworkers provided some data which even suggest fractal surface properties for several carbon blacks [135, 199]. The surfaces of filler particles are not homogeneous. They may be spatially rough, but also energetically heterogeneous. Both effects will enhance the binding of the polymers close to such surfaces. We propose

Fig. 9.1. A cluster of filler particles embedded in the rubber matrix. The cluster itself contributes to the so-called hydrodynamic reinforcement, which is mainly given by the volume effects and the cluster structure. Aggregates with a rough (spatially or energetically) surface contribute significantly to the reinforcement.

Fig. 9.2. Sketch of the main mechanism of adsorption enhancement due to surface roughness: the number of possible binding sites increases without being balanced by a loss in configurational entropy.

that it is always easier to adsorb a polymer on rough surfaces. This can be seen by a simple scaling argument, which is depicted in Fig. 9.2.

9.1.1 Flat surface

First we briefly review a simple scaling treatment of an ideal chain adsorbed on a flat surface, which was introduced by de Gennes [63]. Let $R_\perp$ and $R_\parallel \simeq R_0 \simeq bN^{1/2}$ be the mean sizes of an ideal polymer (with N monomers and effective monomer length b) perpendicular and parallel to the surface, respectively. The monomer density is assumed to be constant in a region of size $R_\perp R_\parallel^2$. Then the number $\mathcal{N}$ of monomers bound to the surface is estimated as

$$\mathcal{N} = bR_\parallel^2 \frac{N}{R_\perp R_\parallel^2} = \frac{bN}{R_\perp}. \tag{9.1}$$

Consequently the free energy can be written as

$$\beta F \approx \frac{R_0^2}{R_\perp^2} - \beta w \mathcal{N} = \frac{b^2 N}{R_\perp^2} - \beta w \frac{bN}{R_\perp}, \tag{9.2}$$

the first term is the confinement energy, and the second one is due to contact interactions with the surface ($-w$ is the attractive monomer attraction, β the inverse temperature so that the product β_w is dimensionless). Minimization of the free energy, $\partial F / \partial R_\perp = 0$, gives an expression for the polymer thickness $R_\perp$ perpendicular to the surface:

$$R_\perp \simeq \frac{b}{\beta w} . \tag{9.3}$$

Thus the thickness of the polymer reduces with growing attractive interaction strength, as expected. The independence of the chain length N indicates that the polymer is in the so-called "localized" regime [200].

9.1.2 Generalization for fractal surfaces

A fractal surface may be characterized by its fractal dimension d_S ($2 \leq d_S \leq 3$), where $d_S = 2$ corresponds to a flat surface. Two examples are shown in Fig. 9.3. The limit $d_S \to 3$ produces an extremely rough, space-filling surface. Brownian surfaces [201] are characterized by $d_S = 2.5$.

Now the number of bounded monomers is written as

$$\mathcal{N} = b^{3-d_S} R_\perp^{d_S} \frac{N}{R_\perp^3} . \tag{9.4}$$

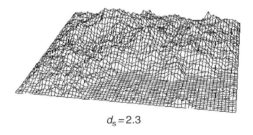

$d_s = 2.3$

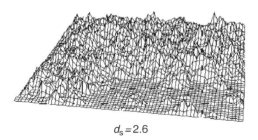

$d_s = 2.6$

Fig. 9.3. The surface roughness can in some cases be measured by the surface fractal dimension d_S.

Running through the same procedure as above yields

$$R_\perp \simeq \frac{b}{(\beta w)^{1/(d_S - 1)}} , \tag{9.5}$$

so that the result (9.3) is recovered for the case of a flat surface, $d_S = 2$.

From (9.3) we have $\beta w < 1$ because $b \ll R_\perp$ for weak adsorption, where no complete collapse on the surface takes place. In fact, for most materials values of βw of ~ 0.01–0.1 are found [202]. Therefore the polymer adsorption on rough surfaces ($d_S > 2$) generally is enhanced compared to the case of a flat surface, i. e. $R_\perp^{\text{rough}} < R_\perp^{\text{flat}}$.

Although this is a crude argument, it gives an insight into the main aspects of adsorption enhancement: the crucial point is the competition between the gain in potential energy obtained by binding to the surface and the loss in chain entropy associated with confined chains in comparison to free chains. Therefore, the dominating feature in our consideration is to have an increasing number of binding sites at a rough surface without paying an increased entropy penalty, which means that a chain has to lose less configurational entropy when adsorbing on a rough surface. This is in agreement with results of much more extensive previous calculations by Douglas [171] and Hone *et al.* [203].

A similar argument holds for the case of energetic heterogeneity [204]: with a distribution of the interaction strength on the surface, the chain can select the strong binding points without changing its configuration too much, thus there is a larger effective interaction strength.

9.2 Variational calculation statics

9.2.1 Variational calculation

For a systematic study of $R_\perp$ in the case of spatial and energetic heterogeneity, the free energy is calculated via a variational procedure, in which the disorder is treated as a quenched (i. e. frozen) randomness. The replica method is avoided by introducing an additional variational parameter, see next section. We consider an ideal chain at an infinite, penetrable, well-defined surface with a low profile. Furthermore, we assume an attractive contact (i. e. extremely short range) interaction between the chain and the surface that can be mimicked by a delta potential.

This system can be represented by its Edwards–Hamiltonian, which reads

$$\beta H = \frac{3}{2b^2} \int_0^N ds \left(\frac{\partial \mathbf{R}(s)}{\partial s}\right)^2 + \beta \int_0^N ds \, V(\mathbf{R}(s)), \tag{9.6}$$

where $\mathbf{R}(s)$ is the chain segment position vector. The potential contains the polymer–surface coupling:

$$V(\mathbf{R}(s)) = -\int d^2x \, K[h(\underline{x})] \, b \, w(\underline{x}) \, \delta(\mathbf{R}(s) - \mathbf{h}(\underline{x})) \,, \tag{9.7}$$

with $\mathbf{h}(\underline{x}) = (\underline{x}, h(\underline{x}))$, where $\underline{x} = (x_1, x_2)$ is an internal surface vector and $h(\underline{x})$ the varying z-component perpendicular to the surface. The surface disorder is described by $w(\underline{x})$ for energetical disorder, i. e. an interaction strength varying on the surface, and by $h(\underline{x})$ for spatial disorder, i. e. a rough surface profile. The factor $K[h(\underline{x})] = (1 + |\nabla h(\underline{x})|^2)^{1/2}$ takes account of the local deflection of the surface in Cartesian coordinates.

In order to approximate the free energy we make use of a Feynman variational procedure:

$$\langle \exp[-\beta(H - H_0)] \rangle_{H_0} \geq \exp[-\beta \langle H - H_0 \rangle_{H_0}], \tag{9.8}$$

where $\langle ... \rangle_{H_0}$ denotes the average with respect to a trial Hamiltonian H_0. This gives an upper bound βF^* to the free energy

$$\beta F \leq \beta F^* = \beta F_0 + \beta \langle H - H_0 \rangle_{H_0} \,, \tag{9.9}$$

where

$$\beta F_0 = \log \left(\int \mathcal{D}\mathbf{R} \exp\{-\beta H_0\} \right) \,. \tag{9.10}$$

βF^* has to be minimized to give the best estimate for the true free energy βF.

9.3 Trial Hamiltonian

An appropriate choice of the trial Hamiltonian is very important when utilizing the variational procedure. Here we take an extension of a form suggested by Garel and Orland [205]:

$$\beta H_0 = \frac{1}{2} \sum_{j=1}^{3} \int_0^N ds \int_0^N ds' \, (R_j(s) - B_j) g_j^{-1}(|s - s'|)(R_j(s') - B_j) \,. \tag{9.11}$$

Its features are: (a) it is quadratic in $\mathbf{R}(s)$, so that an exact calculation of βF_0 is possible; (b) the coupling of chain segments is mediated by the variational parameters $g_j(|s - s'|)$, one for each direction in space: the indices 1 and 2 are identified with the coordinates x_1 and x_2 of the surface parameterization, index 3 corresponds to the z coordinate parallel to the average surface normal; (c) there is an additional variational parameter $\mathbf{B}$, equivalent to a translation of the center of mass of the chain. It should be mentioned that this type of variational principle was originally designed to avoid replica theory in random systems [205]. This is another reason why the Garel–Orland method is chosen here. If the polymer is assumed to stick permanently at some place along the disordered surface, the problem falls into the classes dealing with "quenched disorder" and difficulties arise with replica symmetry breaking. In the following we will show that the Garel–Orland method is indeed useful to treat the problem of polymer adsorption on disordered surfaces as it yields physically sensible results.

Assuming cyclic boundary conditions $\mathbf{R}(N) \equiv \mathbf{R}(0)$, the variational free energy equation (9.9) can now be calculated to give

$$\beta F^* = -\sum_{n=1}^{\infty}\sum_{j=1}^{3}\log\frac{\tilde{g}_j(n)}{b^2} + \sum_{n=1}^{\infty}\sum_{j=1}^{3}N\omega_n^2\frac{\tilde{g}_j(n)}{b^2} + \beta\mathcal{W}(\mathbf{B},\mathbf{G})\,, \qquad (9.12)$$

where $\omega_n = 2n\pi/N$. Here the interaction energy $\mathcal{W}(\mathbf{B},\mathbf{G})$ is the only term that depends on the interaction potential

$$\mathcal{W}(\mathbf{B},\mathbf{G}) = \frac{-Nb}{(2\pi)^{3/2}(G_1 G_2 G_3)^{1/2}}\int d^2x\; K[h(\underline{x})]\,w(\underline{x})$$
$$\exp\left\{-\sum_{i=1}^{3}\frac{(B_i - h_i(\underline{x}))^2}{2G_i}\right\}\,, \qquad (9.13)$$

the $h_i(\underline{x})$ being the components of $\mathbf{h}(\underline{x})$, i.e. $h_3(\underline{x}) \equiv h(\underline{x})$. The parameters G_j are defined by

$$G_j = 2\sum_{n=1}^{\infty}\tilde{g}_j(n) = \frac{2}{N}\sum_{n=1}^{\infty}\int_0^N ds\; \cos(\omega_n s)g_j(s), \qquad j = 1, 2, 3\,. \qquad (9.14)$$

G_j can be identified with the mean square radius of the polymer parallel (G_3) or perpendicular (G_1 and G_2) to the surface normal.

9.3.1 Minimization of the free energy

Following the lines of Garel and Orland, the minimization of βF^* with respect to $\tilde{g}_j(n)$ and $\mathbf{B}$ leads to

$$\nabla_{\mathbf{B}}\mathcal{W}(\mathbf{B},\mathbf{G}) \overset{!}{=} \mathbf{0} \qquad (9.15)$$

and

$$\tilde{g}_j(n) \overset{!}{=} \frac{b^2}{N\omega_n^2 + \beta b^2\dfrac{\partial^2 \mathcal{W}(\mathbf{B},\mathbf{G})}{\partial B_j^2}}\,, \qquad (9.16)$$

because

$$\partial \mathcal{W}(\mathbf{B}, \mathbf{G})/(\partial \tilde{g}_j(n)) = \partial^2 \mathcal{W}(\mathbf{B}, \mathbf{G})/(\partial B_j^2) \,.$$

As discussed by Garel and Orland [205], in general one expects the variational equations to have several solutions. This applies especially to (9.15), since we are considering an infinite surface, e.g. leading to an infinite number of solutions in the case of a periodic surface heterogeneity. All these solutions have equal free energy.

Introducing the notation

$$\alpha_j = \left(\frac{N^2 b^3}{4(2\pi)^{1/2}} \frac{\beta |w_j^{\text{eff}}|}{G_j^{3/2}} \right)^{1/2} , \tag{9.17}$$

the optimized parameter G_j is calculated from (9.16) as

$$G_j = \begin{cases} \dfrac{Nb^2}{4} \dfrac{\coth(\alpha_j) - \alpha_j^{-1}}{\alpha_j} & \text{for } w_j^{\text{eff}} \geq 0 \\[3mm] -\dfrac{Nb^2}{4} \dfrac{\cot(\alpha_j) - \alpha_j^{-1}}{\alpha_j} & \text{for } w_j^{\text{eff}} < 0 \end{cases} . \tag{9.18}$$

The effective interaction strength w_j^{eff} contains all relevant surface and polymer properties and is given by

$$w_j^{\text{eff}} = \frac{(2\pi)^{1/2} G_j^{3/2}}{N b} \left. \frac{\partial^2 \mathcal{W}(\mathbf{B}, \mathbf{G})}{\partial B_j^2} \right|_{\nabla_{\mathbf{B}} \mathcal{W}(\mathbf{B}, \mathbf{G}) = 0} . \tag{9.19}$$

In two special cases results can be obtained very easily:

(i) If there is no interacting surface present, i.e. $w(x) \equiv 0$, then we immediately have $\alpha_j = 0$ and therefore $G_j = Nb^2/12 \equiv R_g^2/2$, and the chain conformation is purely Gaussian in all directions, as expected where R_g is the radius of gyration of the polymer chain, $R_g = Nb^2/6$.

(ii) For an ideal surface, which means $w(x) \equiv w_0$ and $h(\underline{x}) \equiv h_0$, the effective interactions strengths are calculated as $w_{1/2}^{\text{eff}} = 0$ and $w_3^{\text{eff}} = w_0$. So the definition (9.19) of w_j^{eff} guarantees correct results for this case.

The explicit forms of w_j^{eff} for various special sorts of surface heterogeneity are calculated in the next section.

The discussion of (9.18) is complicated by the fact that the effective interaction strength is itself a function of the polymer extensions in different directions. But in general (9.18) can, for $w_j^{\text{eff}} \geq 0$, be expanded in the limits of small and large α_j. This yields the mean polymer extension into the different directions of space, $\bar{R}_3$ being parallel to the surface normal and $\bar{R}_1$ and $\bar{R}_2$ perpendicular to it, if $\langle h(\underline{x}) \rangle = 0$ is assumed:

$$\bar{R}_j \simeq \sqrt{G_j} \simeq \begin{cases} R_0 \left\{ 1 - cN^{1/2}\beta w_j^{\text{eff}} \right\} & \text{for } \beta w_j^{\text{eff}} \ll N^{-1/2} \\[2em] \dfrac{b}{\beta w_j^{\text{eff}}} \left\{ 1 - \dfrac{\pi}{N(\beta w_j^{\text{eff}})^2} \right\} & \text{for } \beta w_j^{\text{eff}} \gg N^{-1/2} \end{cases}, \qquad (9.20)$$

where, as above, R_g denotes the radius of gyration of the corresponding Gaussian chain.

Thus, in the limit of small effective interaction strength the chain has a Gaussian conformation (see Fig. 9.4), whereas for high βw_3^{eff} the chain is localized at the surface, leading to a mean polymer size that in lowest order shows the same characteristics as the result of the scaling argument, (9.3). From the conditions for the limiting cases, a localization criterion,

$$\beta w_{j\,\text{crit}}^{\text{eff}} \approx N^{-1/2}, \qquad (9.21)$$

can be found, which means $\beta w_{j\,\text{crit}}^{\text{eff}} \approx 0$ for long chains. Therefore very long chains are always adsorbed, i. e. localized, at an attractive surface. This, of course, is a consequence of our assumption of a penetrable surface, since in the opposite case of impenetrable surface adsorption takes place only from a finite interaction strength [206], i. e. $\beta w_{j\,\text{crit}}^{\text{eff}} > 0$.

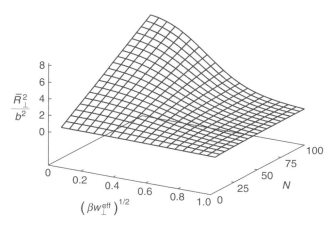

Fig. 9.4. Numerical result for the optimized variational parameter $G_3 \simeq \bar{R}_\perp^2$ as a function of interaction strength $(\beta w_3^{\text{eff}})^{1/2}$ and chain length N. The localization transition can be identified: for small values of w_3^{eff}, G_3 grows linearly with N, whereas G_3 is independent of N for large values of w_3^{eff}.

For a negative effective interaction strength $w_j^{\text{eff}} < 0$, only the case $\beta|w_j^{\text{eff}}| \ll N^{-1/2}$ is important for us, as we are mainly interested in the adsorption behavior. Expansion of (9.18) in this case yields

$$\bar{R}_j \simeq \sqrt{G_j} \simeq R_0 \left\{ 1 + cN^{1/2}\beta w_j^{\text{eff}} \right\} . \tag{9.22}$$

9.3.2 Effective interaction strength

The full general form of the effective interaction strength is

$$w_j^{\text{eff}} = \frac{G_j^{1/2}}{2\pi(G_1 G_2 G_3)^{1/2}} \int d^2x \; K[h(\underline{x})] \; w(\underline{x}) \left(1 - \frac{(B_j - h_j(\underline{x}))^2}{G_j} \right)$$
$$\exp \left\{ -\sum_{i=1}^{3} \frac{(B_i - h_i(\underline{x}))^2}{2G_i} \right\} , \tag{9.23}$$

where the translational parameter $\mathbf{B}$ has to be chosen such that $\nabla_{\mathbf{B}}\mathcal{W}(\mathbf{B}, \mathbf{G}) = \mathbf{0}$. In the following, the surface is assumed to be symmetrical with respect to the coordinates x_1 and x_2. Then we immediately have $B_1 \overset{!}{=} 0$ and $B_2 \overset{!}{=} 0$ as a solution of the minimization equation (9.15). For simplicity, we additionally assume the surface heterogeneity to depend only on one space direction x_1, which means $w(\underline{x}) \equiv w(x_1)$ and $h(\underline{x}) \equiv h(x_1)$. Hence $w_2^{\text{eff}} = 0$ and the polymer extension into the direction of x_2 equals that of a Gaussian chain. In this case the expressions for w_1^{eff} and w_3^{eff} reduce to

$$w_1^{\text{eff}} = \frac{1}{(2\pi G_3)^{1/2}} \int dx \; K[h(x)] \; w(x) \left(1 - \frac{x^2}{G_1} \right)$$
$$\exp \left\{ -\frac{x^2}{2G_1} - \frac{(B_3 - h(x))^2}{2G_3} \right\} , \tag{9.24}$$

$$w_3^{\text{eff}} = \frac{1}{(2\pi G_1)^{1/2}} \int dx \; K[h(x)] \; w(x)$$
$$\exp \left\{ -\frac{x^2}{2G_1} - \frac{(B_3 - h(x))^2}{2G_3} \right\} . \tag{9.25}$$

Now a straightforward calculation for various types of surface heterogeneity is possible.

(1) For a flat surface with energetic heterogeneity, $h(x) = h_0$, the minimization condition (9.15) results in $B_3 \overset{!}{=} 0$, so that the center of mass of the chain is located on the surface. Inserting

$$w(x) = \int_{-\infty}^{\infty} dq \; \exp\{iqx\} \; \tilde{w}(q)$$

leads to

$$w_1^{\text{eff}} = \frac{G_1^{3/2}}{G_3^{1/2}} \int\limits_{-\infty}^{\infty} dq \, q^2 \, \tilde{w}(q) \exp\left\{-\frac{G_1 q^2}{2}\right\} , \tag{9.26}$$

$$w_3^{\text{eff}} = \int\limits_{-\infty}^{\infty} dq \, \tilde{w}(q) \exp\left\{-\frac{G_1 q^2}{2}\right\} . \tag{9.27}$$

As can be seen from the notation $\tilde{w}(q) = w_0 \delta(q) + \tilde{w}^*(q)$, the effective interaction strength parallel to the surface is independent of the mean interaction strength w_0.

- For a periodic interaction strength

$$\tilde{w}(q) = w_0 \delta(q) + (A_w/2)\{\delta(q-f) + \delta(q+f)\}$$

with amplitude A_w and wave number f, we have

$$w_1^{\text{eff}} = \frac{G_1^{3/2}}{G_3^{1/2}} A_w f^2 e^{-G_1 f^2/2} , \tag{9.28}$$

$$w_3^{\text{eff}} = w_0 + A_w e^{-G_1 f^2/2} . \tag{9.29}$$

Thus w_3^{eff} takes on its maximum $w_0 + A_w$ if the wavelength of the heterogeneity exceeds the polymer size parallel to the surface, $\lambda \simeq f^{-1} \gg \bar{R}_1$, because in this case the polymer chain, which is located at a maximum of $w(x)$, does not notice the existence of the minima of the interaction strength. In the opposite case, $f^{-1} \ll \bar{R}_1$, the fluctuations of $w(x)$ can no longer be resolved, w_3^{eff} is minimal and equal to the mean interaction strength. w_1^{eff} is small (leading to a polymer size $\bar{R}_1 \simeq R_0$ parallel to the surface), except when $f^{-1} \approx \bar{R}_1$, and when A_w is large.
- A randomly distributed interaction strength is best handled by identifying the amplitude in Fourier space $\tilde{w}^*(q)$ with the square root of the spectral density $S(q)$, so that

$$\tilde{w}(q) = w_0 \delta(q) + c^{-1} \Delta_w \exp\{-q^2\xi^2\}$$

for a Gaussian distribution with variance Δ_w^2, correlation width ξ, and constant $c = (2\pi)^{1/4}$. Then the effective interaction strengths are

$$w_1^{\text{eff}} = \frac{G_1^{3/2}}{G_3^{1/2}} \frac{c \, \Delta_w}{(G_1 + \xi^2)^{3/2}} , \tag{9.30}$$

$$w_3^{\text{eff}} = w_0 + \frac{c \, \Delta_w}{(G_1 + \xi^2)^{1/2}} . \tag{9.31}$$

The magnitude of the heterogeneity is determined by both ξ and Δ_w: the smaller the correlation width and larger the variance, the stronger are the fluctuations, which leads to an increase of the effective interaction strength. The limiting cases perpendicular to the surface are

$$
w_3^{\text{eff}} \approx
\begin{cases}
w_0 + c\,\Delta_w\,\xi^{-1} & \text{for } \xi \gg \bar{R}_1, \\[2mm]
w_0 + c\,\Delta_w\,\bar{R}_1^{-1} & \text{for } \xi \ll \bar{R}_1.
\end{cases}
\tag{9.32}
$$

(2) In the case of a heterogeneous surface profile (where the interaction strength $w(x) = w_0$ is assumed constant) the disorder has to be weak in order to make the x integration feasible. Therefore we only investigate the case $|h(x)| \ll 1$ and $|\nabla h(x)| \ll 1$, where $\langle h(x) \rangle = 0$, and restrict the calculation to first order in the fluctuation of $h(x)$. With $h(x) = \int_0^\infty dq\, \cos(qx)\, \tilde{h}(q)$, the minimization (9.15) yields $B_3 \approx \int_0^\infty dq\, \tilde{h}(q)\, \exp\{-G_1 q^2/2\}$. This means that the center of mass of the chain to some extent follows the surface profile.

Now the deflection factor can be approximated by

$$
K[h(x)] = (1 + |\nabla h(x)|^2)^{1/2} \approx 1 + \frac{1}{2} \int_0^\infty dq \int_0^\infty dq'\, \tilde{h}(q)\, \tilde{h}(q')\, qq'\, \sin(qx)\, \sin(q'x).
\tag{9.33}
$$

If additionally the part of the exponent in (9.24) and (9.25) that depends on $h(x)$ is expanded, we obtain in lowest order of $\tilde{h}(q)$

$$
\begin{aligned}
w_1^{\text{eff}} \approx w_0 \left(\frac{G_1}{G_3}\right)^{3/2} \int_0^\infty dq \int_0^\infty dq'\, \tilde{h}(q)\, \tilde{h}(q') \exp\left\{-\frac{G_1(q^2 + q'^2)}{2}\right\} \\[2mm]
\times \left[q^2 + \frac{(q + q')^2}{2} \left(G_3 qq' \sinh(G_1 qq') - \cosh(G_1 qq')\right) \right],
\end{aligned}
\tag{9.34}
$$

$$
\begin{aligned}
w_3^{\text{eff}} \approx w_0 \left\{ 1 + \frac{1}{2G_3} \int_0^\infty dq \int_0^\infty dq'\, \tilde{h}(q)\, \tilde{h}(q') \exp\left\{-\frac{G_1(q^2 + q'^2)}{2}\right\} \right. \\[2mm]
\left. \times \left[3 + G_3 qq' \sinh(G_1 qq') - \cosh(G_1 qq') \right] \right\}.
\end{aligned}
\tag{9.35}
$$

- A periodic surface geometry $\tilde{h}(q) = A_h \delta(q - f)$ leads to

$$
\begin{aligned}
w_1^{\text{eff}} \approx w_0 \left(\frac{G_1}{G_3}\right)^{3/2} \frac{A_h^2 f^2}{2} \left\{ G_3 f^2 \left(1 - \exp(-2G_1 f^2)\right) \right. \\[2mm]
\left. - \left(1 - \exp(-G_1 f^2)\right)^2 \right\},
\end{aligned}
\tag{9.36}
$$

$$
\begin{aligned}
w_3^{\text{eff}} \approx w_0 \left\{ 1 + \frac{A_h^2 f^2}{4} \left(1 - \exp(-2G_1 f^2)\right) \right. \\[2mm]
\left. - \frac{3 A_h^2}{4 G_3} \left(1 - \exp(-G_1 f^2)\right)^2 \right\}.
\end{aligned}
\tag{9.37}
$$

For a flat surface, the polymer extension $\bar{R}_3$ parallel to the mean surface normal (i.e. in the z-direction) is identical to the size perpendicular to the surface. This is different for a rough surface profile, which now turns out to be important for the interpretation of (9.36) and (9.37). If, for example, G_3 is small compared with the squared wavelength $\lambda^2 \simeq f^2$, then the polymer sticks to the surface, following the deflections. Therefore the result for $\bar{R}_3 \simeq \sqrt{G_3}$ exceeds the polymer size perpendicular to the surface by the size of the deflection, the effective interaction strength w_3^{eff} is accordingly smaller than w_0. Adsorption enhancement therefore can only be obtained in the opposite case $\bar{R}_3 \gg f^{-1}$, where the effective interaction strength parallel to the surface normal takes on the maximum value

$$w_{3\,\text{max}}^{\text{eff}} = w_0 \{1 + A_h^2 f^2/4\}.$$

Here we have used $G_1 \leq G_3$, which results from the fact that w_1^{eff} is always small according to the condition $A_h f \ll 1$. As can be seen from (9.36), in the case $G_3 \ll f^{-1}$ discussed above, w_1^{eff} is negative, so the mean polymer extension in the x-direction exceeds the size of a corresponding Gaussian chain, see (9.22).

• Similarly to the case of a randomly distributed interaction strength, the amplitude in Fourier space $\tilde{w}^*(q)$ of a randomly distributed surface profile is identified with the square root of the spectral density $S(q)$. This means $\tilde{h}(q) = c^{-1} \Delta_h \exp\{-q^2 \xi^2/2\}$ for a Gaussian distribution with mean 0, variance Δ_h^2, and correlation width ξ. In order to satisfy the requirement of weak disorder, we have to assume $\Delta_h^2 \ll 1$ and $\xi \gg 0$. Then the result for the effective interaction strength in the z-direction is

$$w_3^{\text{eff}} = w_0 \left\{ 1 + \sqrt{\frac{\pi}{2}} \frac{\Delta_h^2}{4} \left[\frac{G_1}{\xi^3} (2G_1 + \xi^2)^{-3/2} \right. \right.$$

$$\left. \left. + \frac{3}{G_3} \left((G_1 + \xi^2)^{-1} - \frac{1}{\xi} (2G_1 + \xi^2)^{-1/2} \right) \right] \right\}. \tag{9.38}$$

For a very large correlation width, which in the limit $\xi \to \infty$ corresponds to a flat surface, we again have the effect of a reduction of the effective interaction strength compared with the flat surface, $w_3^{\text{eff}} < w_0$. Therefore the result relevant for adsorption enhancement is here obtained in the case $\Delta_h \leq \xi^2 \ll G_3 \leq G_1$, where the effective interaction strength has its maximum value,

$$w_{3\,\text{max}}^{\text{eff}} \approx w_0 \left\{ 1 + \frac{c \Delta_h^2}{\xi^3 \sqrt{G_1}} \right\}. \tag{9.39}$$

A closed expression for w_1^{eff} is not available, but the main features of the result can be estimated to strongly resemble those of w_1^{eff} for a periodic surface profile discussed above.

9.4 Some further remarks on the interpretation

The variational calculation presented here is valid for weak spatial disorder only (therefore it does not reproduce the scaling behavior for fractal surfaces). Nevertheless, the mechanism of adsorption enhancement is well reproduced, and we find agreement with the results of all the special cases which have already been investigated in the literature.

A special feature of the variational method employed here is the possibility of quantifying the localization transition, i. e. the transition from a slightly deformed Gaussian coil to a localized conformation, where the polymer size perpendicular to the surface no longer depends on the chain length. According to (9.21) the localization can be obtained by increasing either the effective interaction strength or the chain length. This helps to compare the strength of adsorption enhancement for the two sorts of disorder considered here: as can be seen from the maximum values of (9.28) and (9.37) or from a comparison of (9.30) and (9.39), the localization transition is only slightly affected by a rough surface profile, whereas energetic heterogeneity can induce the transition even at vanishing mean interaction strength. Therefore we conclude that the disorder-induced enhancement of polymer adsorption is much more significant for a heterogeneous interaction strength than for spatial roughness.

Our findings concerning the localization behavior are affected qualitatively by the assumption of surface penetrability: for infinitely long chains at a flat and homogeneous impenetrable surface, the localization transition, in contrast to (9.21), only occurs for some non-zero value of the attractive potential [206]. Nevertheless, we expect our main statement on the significance of adsorption enhancement also to hold for impenetrable surfaces, since the comparison of transparent and opaque surfaces in simple solvable cases by Hone *et al.* [203] shows that they should not be affected differently by weak surface heterogeneities.

The length of the polymer parallel to the surface does not directly depend on the mean interaction strength, but only through the extension perpendicular to the surface. Thus, because it is less affected by heterogeneity, the former always exceeds the latter, except for one special case: for a flat neutral surface with a periodic interaction strength, the length of the polymer parallel to the surface is smaller than perpendicular to it if the period fits the polymer size such that it is concentrated to a maximum of $w(x)$ and even restricted by the neighboring repulsive regions.

Of course, the dynamics is missing and we do not know whether the chains are really localized, i. e. if they are dynamically bound on the surface. We are going to investigate this in the next section.

9.4.1 Modeling by random potentials

For the static problem we used a model to approximate the filler surface in the most appropriate way that allows a theoretical treatment. Further simplifications are needed to study the problem dynamically. A dynamical study is necessary to predict essential contributions to the modulus from localized chains. It turns out that one possibility is the introduction of a random field of certain properties.

The main idea is to reduce and to model the polymer–filler surface interaction in an appropriate way. It turns out that it is most useful to model the filler surface by a random potential. Indeed, this covers the two aspects already mentioned, i. e. spatially rough surfaces and energetically heterogeneous surface. The surface is depicted in Fig. 9.5. The theoretical description starts with a Hamiltonian for the polymer which interacts with the random surface:

$$\beta H = \frac{3}{2b^2} \int_0^N \mathrm{d}n \left(\frac{\partial \mathbf{R}(s)}{\partial s} \right)^2 + \beta \int_0^N \mathrm{d}n\, V(\mathbf{R}(s)). \tag{9.40}$$

Here the first term is nothing but the Gaussian connectivity of the chain, which leads to the Gaussian distribution if the potential is set to be zero. The parameter n counts the monomers along the contour and N is the chain length. The totally random potential $V(\mathbf{R}(s))$ has the properties

$$\langle V(\mathbf{R}(s) \rangle = 0,$$
$$\langle V(\mathbf{R}(s)) V(\mathbf{R}(s')) \rangle = \Delta \delta(\mathbf{R}(s) - \mathbf{R}(s')), \tag{9.41}$$

where the inverse temperature β is now absorbed into the disorder strength Δ.

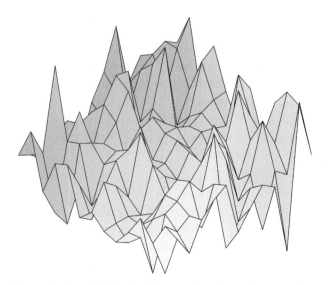

Fig. 9.5. Theoretical model for the filler surface, based on a random potential.

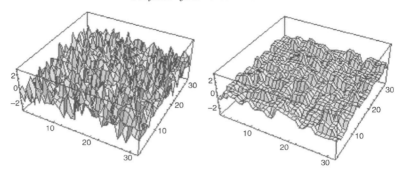

Fig. 9.6. Brownian surfaces of different surface dimensions d_s (see also Fig. 9.3). The closer the surface fractal dimension d_s is to 2, the flatter the shape of the surface.

At first sight such uncorrelated random potentials might not appear appropriate to model the surface of filler particles, but it turns out that they satisfy many conditions which yield the main physical properties in both statics and dynamics. Most of the surface properties are described by the parameter Δ, which denotes the typical volume (of a "hole" in surface).

However, it is not a major problem to generalize all that follows to potentials with different (fractal properties) as shown in Fig. 9.6. Then the additional fractal correlations show up in a surface fractal dimension d_S and the correlation of the potential becomes

$$\langle V(\mathbf{R}(s))V(\mathbf{R}(s'))\rangle = \frac{\Delta b^{-d_S}}{|\mathbf{R}(s) - \mathbf{R}(s')|^{3-d_S}}, \tag{9.42}$$

with its well-known limits: $d_S \to 0$ corresponds to the totally random case since the correlation is very short-range $d_S = 2$ for a completely flat surface, and finally the most interesting case $2 < d_S < 3$ for Brownian surfaces. For the present discussion we will stay, however, with the simplest case, i. e. uncorrelated surfaces.

To see that this random field is sufficient to rederive the previous results we estimate the main effect of the random potential, whose typical barrier can be estimated to be $\sqrt{\langle V(\mathbf{R})V(\mathbf{0})\rangle} \simeq \sqrt{\Delta}N/R^{d/2}$. The main theoretical results can then be put together and summarized in terms of the free energy:

$$\beta F \simeq \frac{R^2}{Nb^2} + \frac{Nb^2}{R^2} - \Delta^{1/2}\frac{N}{R^{3/2}}. \tag{9.43}$$

The first two terms represent the nature of the Gaussian chain in extension and confinement. The negative sign of the effective disorder potential has its origin in the effective attraction from the disorder. In earlier publications it has already been shown that any disorder induces an effective attraction [179, 207]. The special form of the potential corresponds to a typical energy barrier produced by the disorder. This attractive nature is of special significance, since it confirms earlier statements and confines the chain. In order to find a significant chain confinement the disorder Δ must be larger than the entropy term. This yields the condition that the critical surface roughness is $\Delta_{\text{crit}} \simeq b^3 N^{-1/4}$. Only when Δ is larger than this critical value is the filler particle active enough to attract the chains. If this is the case the chains become "localized." Indeed, upon minimization of the free energy we find for the chain size

$$R \simeq b \frac{b^3}{\Delta - \Delta_{\text{crit}}} \approx b \frac{b^3}{\Delta}, \tag{9.44}$$

which means that the size of the chain is entirely determined by the disorder (the size of the "holes" or valleys in Fig. 9.7). Note that the result agrees basically with those derived previously, since here we have found a result for chain size independent of the chain length. Physically this shows that the chain is attracted by the surface and localized in an appropriate hole of typical size as shown in Fig. 9.7. It is important to realize that the chain conformations are totally determined by the disorder, thus upon localization the chain adopts the disorder size. So far the arguments have applied to a single chain. In filled rubbers the chains are not free but are bound into a network. However, it can be shown that similar arguments apply. The only change which has to be made concerns the disorder strength. It can be shown by a simple calculation that in the case of networks the localization criterion modifies to

$$\frac{\Delta}{b^3} > N_{\text{mesh}} \underset{\text{dense networks}}{\Longrightarrow} \Delta > \xi^3, \tag{9.45}$$

where N_{mesh} is the mesh size of the network and ξ is the corresponding correlation length, i.e. the mean distance between crosslinks (see Fig. 9.8). More detailed

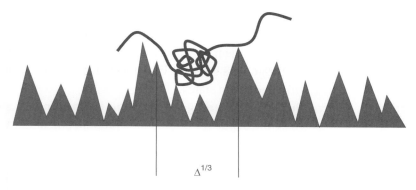

$\Delta^{1/3}$

Fig. 9.7. Localization of a chain in a typical spatial or energetic hole of size Δ.

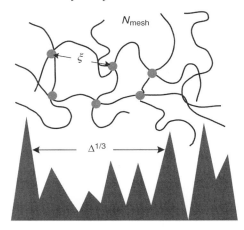

Fig. 9.8. Localization of network chains in a typical spatial or energetic hole of size Δ.

calculations show that for a given (sufficiently large) filler activity Δ, parts of the network localize in a way similar to that for the free chain.

9.4.2 Annealed and quenched disorder

Let us consider first the case in which the characteristic times of the chain's configurations and the chain's center-of-mass positions in a disordered medium are of the same order of magnitude. Then in the course of an experiment the chain experiences all possible quenched field realizations. This corresponds to annealed disorder with the corresponding free energy $F_{\text{anneal}} = -\ln \langle \Xi \rangle_V$, where Ξ is the partition function at a given realization of $V(\mathbf{r})$ and $\langle \cdots \rangle_V$ is the averaging over the field $V(\mathbf{r})$ distribution. It can be shown that in this case

$$\langle \Xi \rangle_V = \Xi_0\{v - \Delta\}, \tag{9.46}$$

where $\Xi_0\{v\}$ is the partition function of the pure (i. e. without disorder) system with the second virial coefficient v. As a result the only effect of the disorder is the reduction of the second virial coefficient, i. e. a reduction of the excluded volume. However, the disorder may change the sign of the effective second virial coefficient, and therefore can cause collapsed chain states.

Nevertheless, in a medium with a strong disorder the chain is preferentially trapped in some regions where the depth of the quenched random potential exceeds $k_B T$. In this case the chain is pinned down in some particular place in a disordered medium and experiences only a local quenched field. This corresponds to quenched disorder and the relevant free energy is $F_{\text{quench}} = -\langle \ln \Xi \rangle_V$. For the site-diluted lattice medium model it was argued by Machta [208] and Machta and Kirkpatrik

[209] that, while the size of the chain is unaffected by the disorder (i. e. $R \sim bN^{\nu}$), the whole spatial distribution of the chain is correlated with the disorder for $d < 4$. The effect of the disorder shows up as an essential singularity in the N dependence of the so-called typical value of the partition function, $\Xi_{\text{typ}} = \exp\left(-F_{\text{quench}}\right)$, as well as the chain's center-of-mass diffusion coefficient D.

9.4.3 Dynamics of localized chains – freezing, glass transition at filler surfaces

So far we have discussed only the static picture. It is very challenging to study the dynamics of localized chains and to discuss their contribution to the viscoelastic properties of reinforced elastomers. These factors are of special importance since they enable us to examine the nature of polymer dynamics in the localized phase. Moreover, they describe the dynamic behavior of the bound "rubber phase" and the change in the local dynamics of the chains, which will contribute to the shear modulus $G(\omega)$ in a natural way.

We have studied the dynamics of polymers confined in a random potential by using Langevin dynamics. This is the natural way to find modifications of the motion of the center of mass and the change of the Rouse modes. In the following we will only summarize the results and leave the details of the calculations to elsewhere.

The first observation is that the center-of-mass (CM) diffusion "freezes," and we have

$$D_{\text{CM}} = D_0(1 - \Delta/\Delta_{\text{crit}}), \tag{9.47}$$

where D_0 denotes the bare diffusion constant without disorder. This result has been confirmed by numerical simulations [271]. Therefore, we may conclude that the chains are dynamically localized as well. Once the filler activity exceeds a certain value Δ_{crit}, the chains become frozen in the disorder, i. e. they no longer diffuse and are bound by the surface. This result suggests a naive estimate of the shift of the glass transition temperature of the chains localized at the active surface $\Delta T_g \propto \left(\Delta/b^3\right)^2$.

A more important problem concerns the internal modes of the chain. If the chains freeze in the disordered surfaces, several internal modes have to freeze also. It is not sufficient that the center-of-mass motion ceases, several Rouse modes must also freeze out such that the chain can localize. To do so we use the simplest approximation and study Rouse chains in the random potential. Usually the chain dynamics is described by Rouse modes, which decompose the chain motion into different modes which are characterized by a typical relaxation time

$$\tau_q \propto \frac{\zeta N^2 b^2}{k_{\text{B}} T} \frac{1}{q^2}. \tag{9.48}$$

Here ζ is the friction coefficient and q represents the Rouse modes. Large values of q correspond to small distances in the chain, i. e. local motions inside the chain. Small values of q correspond to large-scale motions. The center-of-mass motion can be viewed as the limit $q \to 0$. Now let us see what may happen with chain modes. To do so, we can formally use the idea of decomposing the motion of the chain into Rouse modes. Therefore, we use the mode version of the chain Hamiltonian, which will provide a (too) simple view of the problem. Using Rouse modes, the chain Hamiltonian reads

$$\beta H = \frac{d}{2b^2} \sum_q q^2 |\mathbf{R}_q|^2 - \sqrt{\Delta} \frac{N}{R^{d/2}}, \tag{9.49}$$

which defines a proper Langevin equation of the form

$$\zeta \frac{\partial \mathbf{R}_q}{\partial t} = -\frac{\delta H}{\delta \mathbf{R}_q} + \mathbf{f}_q(t). \tag{9.50}$$

Then we may try a simple approximation which corresponds to the case of mode decoupling and use Harris's estimate for the disorder term:

$$\sqrt{\Delta} \frac{N}{R^{d/2}} \sim \sqrt{\Delta} N^{(1-\nu d/2)}. \tag{9.51}$$

We know, however, that for larger values of the disorder Δ the chain becomes localized, which means, that the chain size does not depend on the chain length, i. e. $R \sim N^0$, which yields immediately

$$\left\langle |\mathbf{R}_q|^2 \right\rangle \sim \frac{1}{q}. \tag{9.52}$$

If we put (9.52) back into the Hamiltonian (9.49) we find a rough criterion for a critical localization mode q_{crit}, which depends on the disorder and chain length according to:

$$q_{\mathrm{crit}}^2 < \Delta N^{(2-\nu d)}, \tag{9.53}$$

so that all large-scale motions are frozen. Although these preliminary answers seem to be plausible, the situation is far more complicated. The main reasons for this arise from the coupling of the chain modes due to the interactions that are induced by the irregular nature of the filler surfaces.

What can we expect? Usually in free polymer chains all motions relax to zero, i. e. their correlation function can be described by

$$\text{correlations} \propto \exp(-t/\tau_q). \tag{9.54}$$

In localized Rouse chains this no longer holds. It can be shown that disorder-induced freezing of the chains implies a non-exponential decay. Moreover, certain correlations no longer relax to zero and instead follow a law like

$$\text{correlations} \propto \exp\left(-(t/\tau_q)^\beta\right) + f(q). \tag{9.55}$$

The appearance of the non-ergodicity parameter $f(q)$, which is mode-dependent, reflects the localization process. Indeed, if the non-ergodicity parameter is zero, all relaxations should be exponential, i. e. $\beta = 1$. We expect therefore a non-zero value for $f(q)$ from a certain value of the filler activity or disorder parameter Δ. The resulting general theory of mode coupling can be summarized by the following simple equation:

$$q^4 < \rho \Delta N = q_{\text{crit}}, \tag{9.56}$$

which is interpreted as follows. First, all modes which require the inequality are frozen. In general, this means that large-scale motions freeze out at a certain disorder. The criterion for the freezing is also determined by the radius of gyration, $R_g = bN^\nu$, and the chain density, $\rho = N/R^3$. This means that only the appropriate chain density with its N Rouse modes fits into the "hole" of volume Δ and the chain localizes its modes which satisfy the inequality (9.56). Then $q < q_{\text{crit}}$ localized modes are fitted inside, and only local motion is possible. A more general analysis will follow. Here we just summarize the results in an intuitive figure, Fig. 9.9.

9.5 Equation of motion for the time correlation function

9.5.1 Langevin dynamics

In this section we consider the Langevin dynamics of a polymer chain in the quenched random field. The dynamics of the chain is described by the following Langevin equation:

| Unfavorable region | Chain volume R^3 and disorder Δ do not fit | Disorder defines size- only modes inside Δ move |

Fig. 9.9. Three different situations for a chain close to a disordered surface. The chain localizes inside a "disorder hole," when size and modes fit. A more detailed analysis using a mode-coupling theory follows in the next section.

$$\zeta_0 \frac{\partial}{\partial t} R_j(s,t) - \epsilon \Delta_s R_j(s,t) + \frac{\delta}{\delta R_j(s,t)} H_{\text{int}} \{\mathbf{R}(s,t)\}$$

$$+ \frac{\delta}{\delta R_j(s,t)} V\{\mathbf{R}(s,t)\} = f_j(s,t), \tag{9.57}$$

where j labels the Cartesian components. To be more specific we use the variable s as a discrete variable, ζ_0 is a bare friction coefficient and the second-order finite difference is given by

$$\Delta_s R_j(s,t) = R_j(s+1,t) + R_j(s-1,t) - 2R_j(s,t).$$

The Langevin problem in question becomes much more convenient for theoretical investigation if we change to the Martin–Siggia–Rose (MSR) generating functional representation [210]. The generating functional (GF) of our problem can be written as

$$Z\{\cdots\} = \int \mathrm{D}R_j(s,t) \mathrm{D}\hat{R}_j(s,t) \exp \left\{ A_{\text{intra}} \left[\mathbf{R}(s,t), \hat{\mathbf{R}}(s,t) \right] \right.$$

$$\left. + A_{\text{ext}} \left[\mathbf{R}(s,t), \hat{\mathbf{R}}(s,t) \right] \right\}, \tag{9.58}$$

where the intrachain action is given by

$$A_{\text{intra}} \left[\mathbf{R}(s,t), \hat{\mathbf{R}}(s,t) \right] = \sum_{s=0}^{N-1} \int \mathrm{d}t \left\{ i\hat{R}_j(s,t) \left[\zeta_0 \frac{\partial}{\partial t} R_j(s,t) - \epsilon \Delta_s R_j(s,t) \right] \right.$$

$$\left. + \frac{\delta}{\delta R_j(s,t)} H_{\text{int}} \{R_j(s,t)\} + k_B T \zeta_0 \left[i\hat{R}_j(s,t) \right]^2 \right\} \tag{9.59}$$

and the action related with the quenched random field reads

$$A_{\text{ext}} \left[\mathbf{R}(s,t), \hat{\mathbf{R}}(s,t) \right] = \sum_{s=0}^{N-1} \int \mathrm{d}t\, i\hat{R}_j(s,t) \frac{\delta}{\delta R_j(s,t)} V\{\mathbf{R}(s,t)\}, \tag{9.60}$$

where i is the imaginary unit, $i = \sqrt{-1}$. Equations (9.58)–(9.60) correspond to a given realization of the random field $V\{\mathbf{R}(s,t)\}$. Now we perform the averaging over all configurations of $V\{\mathbf{R}(s,t)\}$ taking into account its Gaussian statistics.

The resulting GF takes the following form:

$$\langle Z\{\cdots\}\rangle_V = \int DR_j(s,t)D\hat{R}_j(s,t) \exp\left\{ A_{\text{intra}}\left[\mathbf{R}(s,t),\hat{\mathbf{R}}(s,t)\right]\right.$$

$$+ \Delta \sum_{s=0}^{N-1}\sum_{s'=0}^{N-1} \int dt\,dt' \int \frac{d^d k}{(2\pi)^d} k_j k_l \exp\left\{i\mathbf{k}[\mathbf{R}(s,t)-\mathbf{R}(s',t')]\right\}$$

$$\left. \times i\hat{R}_j(s,t)i\hat{R}_l(s',t')\right\}. \tag{9.61}$$

It can be seen from (9.61) that averaging over the disorder leads to the non-Markovian (i. e. non-local in time) renormalization of the friction coefficient (which is coupled with $i\hat{R}_j(s,t)i\hat{R}_l(s',t')$). This causes dynamical slowing down and ergodicity breaking which we will discuss below.

9.5.2 Self-consistent Hartree approximation

In order to handle the functional integral (9.61), we use the Hartree approximation. In this approximation the full MSR action is replaced by the Gaussian one in such a way that all terms which include more than two fields $\mathbf{R}(s,t)$ and/or $\hat{\mathbf{R}}(s,t)$ are written in all possible ways as products of pairs of $\mathbf{R}(s,t)$ and/or $\hat{\mathbf{R}}(s,t)$ coupled to the self-consistent averages of the remaining fields. On the other hand in [211] it was shown that the Hartree approximation is equivalent to taking into account Gaussian fluctuations around the saddle-point solution. The resulting Hartree action is a Gaussian functional with coefficients which could be represented in terms of correlation and response functions. The calculation of these coefficients is straightforward and details can be found in the Appendix B of [212]. The second and third virial terms in $A_{\text{intra}}[\mathbf{R}(s,t),\hat{\mathbf{R}}(s,t)]$, as well as the term which is responsible for the non-Markovian renormalization of the friction coefficient, are treated in the same manner as in [213]. After collection of all these terms the final Hartree GF reads.

$$\langle Z\{\cdots\}\rangle_V = \int D\mathbf{R}D\hat{\mathbf{R}} \exp\left\{ A_{\text{intra}}^{(0)}[\mathbf{R},\hat{\mathbf{R}}]\right.$$

$$+ \sum_{s=0}^{N-1}\sum_{s'=0}^{N-1} \int_{-\infty}^{\infty} dt \int_{-\infty}^{t} dt'\, i\hat{R}_j(s,t)R_j(s',t')\lambda(s,s';t,t')$$

$$- \sum_{s=0}^{N-1}\sum_{s'=0}^{N-1} \int_{-\infty}^{\infty} dt \int_{-\infty}^{t} dt'\, i\hat{R}_j(s,t)R_j(s,t)\lambda(s,s';t,t')$$

$$\left. + \frac{1}{2}\sum_{s=0}^{N-1}\sum_{s'=0}^{N-1} \int_{-\infty}^{\infty} dt \int_{-\infty}^{\infty} dt'\, i\hat{R}_j(s,t)i\hat{R}_j(s',t')\chi(s,s';t,t')\right\}, \tag{9.62}$$

where

$$\lambda(s, s'; t, t') = \frac{\Delta}{d} G(s, s'; t, t') \int \frac{\mathrm{d}^d k}{(2\pi)^d} k^4 F(\mathbf{k}; s, s'; t, t')$$

$$+ \int \frac{\mathrm{d}^d k}{(2\pi)^d} k^2 v(\mathbf{k}) F_{\mathrm{st}}(\mathbf{k}; s, s')$$

$$+ \sum_{s''=1}^{N} \int \frac{\mathrm{d}^d k \mathrm{d}^d q}{(2\pi)^{2d}} k^2 w(\mathbf{k}, \mathbf{q}) F_{\mathrm{st}}(\mathbf{q}; s', s'') F_{\mathrm{st}}(\mathbf{k}; s, s') \quad (9.63)$$

and

$$\chi(s, s'; t, t') = \Delta \int \frac{\mathrm{d}^d k}{(2\pi)^d} k^2 F_{\mathrm{st}}(\mathbf{k}; s, s'). \tag{9.64}$$

In (9.63)–(9.64) the response function is

$$G(s, s'; t, t') = \left\langle i\hat{\mathbf{R}}(s', t') \mathbf{R}(s, t) \right\rangle \tag{9.65}$$

and the chain density correlator is

$$F(\mathbf{k}; s, s'; t, t') = \exp\left\{ -\frac{k^2}{d} Q(s, s'; t, t') \right\}, \tag{9.66}$$

with

$$Q(s, s'; t, t') \equiv \langle \mathbf{R}(s, t) \mathbf{R}(s, t) \rangle - \left\langle \mathbf{R}(s, t) \mathbf{R}(s', t') \right\rangle, \tag{9.67}$$

while $F_{\mathrm{st}}(\mathbf{k}; s, s')$ denotes the static limit of (9.66). The angle brackets denote the self-consistent averaging with the Hartree GF which is given by (9.62).

In general, one should consider fluctuation dissipation theorem (FDT) violation which is well known in the context of the glass transition phenomenon [214]. Here we are mainly interested in freezing conditions as well as anomalous diffusion at relatively short times. This enables us to assume that the FDT and the time translational invariance (TTI) are valid, then

$$G(s, s'; t - t') = (k_{\mathrm{B}} T)^{-1} \frac{\partial}{\partial t'} Q(s, s'; t - t') \quad \text{at } t > t'. \tag{9.68}$$

By employing (9.68) in (9.62)–(9.64) and after integrating by parts with respect to the time argument t', we obtain the following Hartree GF:

$$
\langle Z\{\cdots\}\rangle_V = \int \mathbf{DR}\mathbf{D}\hat{\mathbf{R}} \exp\left\{ \sum_{s,s'=0}^{N-1} \int_{-\infty}^{\infty} dt \int_{-\infty}^{t} dt'\, i\hat{R}_j(s,t) \right.
$$

$$
\times \left[\zeta_0 \delta(t-t') + \theta(t-t')\Gamma(s,s';t,t') \right] \frac{\partial}{\partial t} R_j(s',t')
$$

$$
- \sum_{s,s'=0}^{N-1} \int_{-\infty}^{\infty} dt \int_{-\infty}^{t} dt'\, i\hat{R}_j(s,t)\, \Omega(s,s')\, R_j(s',t)
$$

$$
+ \sum_{s,s'=0}^{N-1} \int_{-\infty}^{\infty} dt \int_{-\infty}^{\infty} dt'\, i\hat{R}_j(s,t)
$$

$$
\left. \left[\zeta_0 \delta(t-t') + \theta(t-t')\Gamma(s,s';t,t') \right] i\hat{R}_j(s',t') \right\}, \quad (9.69)
$$

where the memory function is

$$
\Gamma(s,s';t,t') = \Delta \int \frac{d^d k}{(2\pi)^d}\, k^2 F(\mathbf{k};s,s';t,t') \quad (9.70)
$$

and the effective elastic susceptibility is

$$
\Omega(s,s') = \epsilon \delta_{ss'} \Delta_s - \int \frac{d^d k}{(2\pi)^d}\, k^2 (v(\mathbf{k}) - \Delta)\left[F_{st}(\mathbf{k};s,s') - \delta_{ss'} \sum_{s''=0}^{N-1} F_{st}(\mathbf{k};s,s'') \right]
$$

$$
- \frac{1}{2} \sum_{s''=0}^{N-1} \int \frac{d^d k\, d^d q}{(2\pi)^{2d}}\, k^2 w(\mathbf{k},\mathbf{q}) \times \left[F_{st}(\mathbf{k};s,s') F_{st}(\mathbf{q};s'',s') \right.
$$

$$
\left. - \delta_{ss'} \sum_{s'''=0}^{N-1} F_{st}(\mathbf{k};s,s''') F_{st}(\mathbf{q};s''',s'') \right]. \quad (9.71)
$$

In (9.69)–(9.71) we use for simplicity the units of $k_B T = 1$, so that the disorder parameter Δ has the dimensionality of volume. The memory function (9.70) is responsible for the non-Markovian renormalization of the Stokes friction coefficient ζ_0, which arises from interaction with the quenched field $V(\mathbf{k})$. The effective elastic susceptibility (9.71) takes into account all non-dissipative (reactive) forces in the system: the local spring interaction, the renormalization of the second virial coefficient due to the random field $V(\mathbf{k})$, as well as the third virial term.

9.5.3 Equation of motion

The equation of motion for the correlation function

$$C(s, s'; t, t') = \langle \mathbf{R}(s, t)\mathbf{R}(s', t') \rangle \tag{9.72}$$

can readily be obtained from GF (9.69). The result at $t > t'$ reads

$$\zeta_0 \frac{\partial}{\partial t} C(s, s'; t, t') - \sum_{m=1}^{N} \Omega(s, m; t) C(m, s'; t, t')$$

$$+ \sum_{m=0}^{N-1} \int_{t'}^{t} \Gamma(s, m; t, \tau) \frac{\partial}{\partial \tau} C(m, s'; \tau, t') d\tau = 0. \tag{9.73}$$

It is convenient to make the Rouse transformation [62]

$$C(p, t) = \frac{1}{N} \sum_{s=0}^{N-1} C(s, t) \exp(isp) \tag{9.74}$$

and

$$C(s, t) = \sum_{p=0}^{2\pi} C(p, t) \exp(-isp), \tag{9.75}$$

where $p = 2\pi j/N$ $(j = 0, 1, \ldots, N-1)$, i. e. for simplicity we have used the cyclic boundary conditions. After this transformation (9.73) is simplified and takes the form

$$\zeta_0 \frac{\partial}{\partial t} C(p; t) + N \int_0^t \Gamma(p, t - t') \frac{\partial}{\partial t'} C(p; t') dt' + \Omega(p, t) C(p; t) = 0, \tag{9.76}$$

where

$$N\Gamma(p, t) = \Delta \frac{d^{\frac{d}{2}+2}}{2^{d+1}\pi^{d/2}} \sum_{n=0}^{N-1} \frac{\cos(ps)}{[Q(n, t)]^{\frac{d}{2}+1}} \tag{9.77}$$

and

$$\Omega(p) = \frac{2d}{b^2}(1 - \cos p) - (v - \Delta) \frac{d^{\frac{d}{2}+2}}{2^{d+1}(\pi)^{\frac{d}{2}}} \sum_{n=0}^{N-1} \frac{1 - \cos(pn)}{[Q_{st}(n)]^{\frac{d+2}{2}}}$$

$$- w \frac{d^{d+2}}{4^{d+1}(\pi)^d} \sum_{n=0}^{N-1} \sum_{m=0}^{N-n-1} \frac{1 - \cos(pn)}{[Q_{st}(n)]^{\frac{d+2}{2}} [Q_{st}(n)]^{\frac{d}{2}}}. \tag{9.78}$$

In (9.77) and (9.78) the time-dependent mean-square distance,

$$Q(s,t) = \tfrac{1}{2} \left\langle [\mathbf{R}(s,t) - \mathbf{R}(0,0)]^2 \right\rangle,$$

$$= \sum_{p=0}^{2\pi} [C_{st}(p) - \cos(ps)C(p,t)], \tag{9.79}$$

and its static limit,

$$Q_{st}(s) = \sum_{p=0}^{2\pi} [1 - \cos(ps)] \, C_{st}(p), \tag{9.80}$$

make the whole equation of motion for $C(p,t)$ self-consistently closed. In the course of deriving (9.76)–(9.78) we took into account that the segment–segment interaction is short range, i. e. $v(\mathbf{k}) \approx v$ and $w(\mathbf{k}, \mathbf{q}) \approx w$. We have also used the Rouse transformation of the chain density correlator, i. e.

$$F(\mathbf{k}; p; t) = \frac{1}{N} \sum_{n=0}^{N-1} \cos(pn) \exp\left\{ -\frac{k^2}{d} Q(n,t) \right\}. \tag{9.81}$$

The static limit (i. e. $t \to 0$) is evident from (9.76) provided that the initial condition [210]

$$\zeta_0 \left(\frac{\partial}{\partial t} C(p;t) \right)_{t \to 0^+} = \zeta_0 G(p, t \to 0^+) = -\frac{d}{N} \tag{9.82}$$

is taken into account. Then the static equation becomes

$$[NC_{st}(p)]^{-1} = \frac{2}{b^2}(1 - \cos p) - (v - \Delta) \frac{d^{\frac{1}{2}d+1}}{2^{d+1}(\pi)^{d/2}} \sum_{n=0}^{N-1} \frac{1 - \cos(pn)}{[Q_{st}(n)]^{(d+2)/2}}$$

$$- w \frac{d^{d+1}}{4^{d+1}(\pi)^d} \sum_{n=0}^{N-1} \sum_{m=0}^{N-n-1} \frac{1 - \cos(pn)}{[Q_{st}(n)]^{(d+2)/2} [Q_{st}(n)]^{d/2}}. \tag{9.83}$$

It is worth mentioning that this equation is very similar to that for other cases and has a quite general physical meaning in the mode coupling theory of interacting polymer chains [215]. Here the excluded volume interaction is just shifted by the disorder strength Δ, i.e., $v \to v - \Delta$. In the static limit this shift is the only consequence of the random field $V(\mathbf{r})$ effect.

9.6 Dynamic behavior of the chain

We are now in position to launch a more elaborate investigation into the chain dynamic behavior which is based on (9.76)–(9.79). There are at least two topics which can be studied: (i) the anomalous diffusion on the interval between a microscopic characteristic time τ_d (see below) and the longest internal relaxation time τ_R [216, 217]; (ii) Rouse modes dynamical freezing at $t \to \infty$.

9.6.1 Anomalous diffusion

The presence of the quenched random field restricts the motion of the chain at the time interval

$$\tau_d < t < \tau_0 N^{1+2\nu}, \tag{9.84}$$

where τ_d is a crossover time at which the disorder starts to manifest (the value of τ_d will be discussed below) and $\tau_0 N^{1+2\nu}$ is the maximal Rouse time [62] with the Flory exponent ν. This restriction manifests itself through the subdiffusional regimes (anomalous diffusion).

Let us start from the general solution of (9.76). For the Laplace correlator

$$C(p, z) = \int_0^\infty dt\, C(p, t) \exp(-zt), \tag{9.85}$$

this solution reads [210]

$$C(p, z) = \frac{C_{\text{st}}}{z + \frac{\Omega(p)}{\zeta_0 + N\Gamma(p, z)}}. \tag{9.86}$$

The calculation of $\Gamma(p, z)$ is based on (9.77), where the time-dependent mean-square distance $Q(s, t)$ at the time interval (9.84) is approximated by

$$Q(s, t) = b^2 \left(\frac{t}{\tau_0}\right)^{2\theta} + Q_{\text{st}}(s), \tag{9.87}$$

with $\theta = \nu/(1 + 2\nu)$ and $Q_{\text{st}}(s) = b^2 s^{2\nu}$. This form can be justified by implementing simple scaling arguments for a pure (i. e. without disorder) model. The substitution of (9.87) into (9.77) leads to the following result:

$$N\Gamma(p \to 0, t) = A \left(\frac{\Delta}{b^{d+2}}\right) \left(\frac{\tau_0}{t}\right)^\beta, \tag{9.88}$$

with

$$
A = \frac{d^{\frac{d}{2}+2}\tilde{\Gamma}\left(\frac{1}{2\nu}\right)\tilde{\Gamma}\left(\frac{d}{2}-\frac{1}{2\nu}+1\right)}{2^{d+2}\pi^{\frac{d}{2}}\nu\tilde{\Gamma}\left(\frac{d}{2}+1\right)},
\tag{9.89}
$$

where $\tilde{\Gamma}(x)$ is the gamma function and

$$
\beta = \theta\left(d+2-\frac{1}{\nu}\right)
$$

$$
= 1 - \frac{\alpha}{2\nu+1} < 1,
\tag{9.90}
$$

in which $\alpha = 2 - \nu d$ is the "specific-heat" exponent.

The Laplace transformation of (9.88) at $\tau_0 z \ll 1$ reads

$$
N\Gamma(p \to 0, t) = A\left(\frac{\Delta}{b^{d+2}}\right)^{\beta}\tau_0^{\beta}\left(\frac{1}{z}\right)^{1-\beta}.
\tag{9.91}
$$

When the memory term exceeds the bare friction coefficient, i. e. at $t > \tau_d$, we can use (9.91) in (9.86), which after inverse Laplace transformation can be put in the form

$$
C(p,t) = C_{\text{st}}(p)\sum_{k=0}^{\infty}\frac{\left[-\left(\frac{b^{d+2}\Omega(p)}{\Delta A}\right)\left(\frac{t}{\tau_0}\right)^{\beta}\right]^k}{\tilde{\Gamma}(k\beta+1)}.
\tag{9.92}
$$

The center-of-mass mean-square displacement is given by

$$
Q_{\text{CM}}(t) = \frac{1}{2}\left\langle[\mathbf{R}_{\text{CM}}(t) - \mathbf{R}_{\text{CM}}(0)]^2\right\rangle
$$

$$
= \lim_{p \to 0}\{C_{\text{st}}(p) - C(p,t)\}.
\tag{9.93}
$$

Substituting (9.92) into (9.93) results in the leading term of the anomalous diffusion, i. e.

$$
Q_{\text{CM}}(t) = \frac{\mathcal{D}_0}{N}\left(\frac{t}{\tau_0}\right)^{\beta},
\tag{9.94}
$$

where

$$
\mathcal{D}_0 = \frac{b^{d+2}}{\Delta A}.
\tag{9.95}
$$

In the course of deriving (9.94) we used the static equation (9.83), i. e. $C_{\text{st}}(p)\Omega(p) = 1/N$.

It is easy now to estimate the crossover time τ_d after which the disorder starts to affect the diffusion (see (9.84)). The condition for that, $\zeta_0 = \int_0^{\tau_d} dt\, \Gamma(p \to 0, t)$, can be recast in a form

$$\tau_d = \left(\frac{b^{d+2} \, \zeta_0}{\Delta \, A \, \tau_0^{\beta}} \right)^{1/(1-\beta)}. \tag{9.96}$$

One can see that the anomalous diffusion exponent β does not depend on the strength of disorder, whereas the prefactor $\mathcal{D}_0$ decreases with increasing Δ. For a chain in a good solvent, $\nu = 3/(d+2)$ and at $d = 3$ the exponent $\beta_{SAW} \approx 0.9$. For a Gaussian chain $\nu = 1/2$ and $\beta_{Gauss} = 0.75$, i. e. the subdiffusional exponent has the same value as in a polymer melt [210]. Finally in the case of the globule state, $\nu = 1/3$ and $\beta_{Globule} = 0.4$, i. e. globule anomalous diffusion is suppressed by the disorder at most.

9.6.2 Center-of-mass freezing

Let us consider now large-time center-of-mass diffusion. In this case the characteristic time interval

$$t \gg \tau_0 N^{1+2\nu} \tag{9.97}$$

and internal Rouse modes are already relaxed. For this time regime, a reasonable approximation for $Q(n, t)$ has the following form (compare with (9.87)):

$$Q(s, t) = d D t + Q_{st}(s), \tag{9.98}$$

where D is the full (not bare) diffusion coefficient, which is renormalized by the effect of disorder and should be found self-consistently. The equation for the zero-mode diffusion coefficient has the form [210, 212, 218, 219]:

$$D = \frac{1}{N \left[\zeta_0 + N \int_0^{\infty} dt\, \Gamma(p = 0, t) \right]} \tag{9.99}$$

(we recall that in our units of measurement $k_B T = 1$). Equation (9.99) enables us to find D self-consistently. By making use of (9.77) and (9.98) in (9.99) we obtain the following result for the center-of-mass diffusion coefficient

$$D = D_R \left(1 - \Delta \, \mathcal{F}_N \right), \tag{9.100}$$

where $D_R = (\zeta_0 N)^{-1}$ is the Rouse diffusion coefficient and

$$\mathcal{F}_N = \frac{d^{d/2}}{2^d \pi^{d/2}} \, N \sum_{s=0}^{N-1} \frac{1}{[Q_{st}(s)]^{d/2}}. \tag{9.101}$$

It can be seen that at $\Delta \mathcal{F}_N \geq 1$ the center-of-mass diffusion is frozen and the system becomes non-ergodic. The relevance of this result is two-fold. First, this is a particular case of the so-called A-type dynamical phase transition, which has been extensively discussed in the context of the mode coupling theory [220]. Second, if we substitute $Q_{st}(s)$ in (9.101) by its most representative term $Q_{st} \approx b^2 N^{2v}$, we will find $D \approx D_R \left[1 - \mathrm{const}(\Delta/b^d)N^{2-vd}\right]$. As a result we return to Machta's formula or, more exactly, to its expansion up to first order with respect to $(\Delta/b^2)N^{2-vd}$ [208, 209]. This means that (9.100) overestimates the freezing and one should rather treat (9.100) as a cross-over criterion for the weak ergodicity breaking transition in the sense of the results of [214].

9.6.3 Rouse modes freezing and a two mode toy model

Now we are going to study the freezing or the ergodicity breaking of Rouse modes with $p \neq 0$. This phenomenon mathematically manifests itself as a bifurcation with respect to the non-ergodicity function, which, in turn, is a long-time limit of the corresponding correlator [220]. Let us define the persistent part of the normalized correlator (i. e. the non-ergodicity function) as the long-time limit

$$f(p) = \lim_{t \to \infty} \frac{C(p,t)}{C_{st}(p)}. \tag{9.102}$$

The equation for $f(p)$ can easily be obtained by taking the limit $t \to \infty$ in (9.76). The result reads

$$\frac{f(p)}{1 - f(p)} = \Delta \frac{d^{\frac{1}{2}d+1}}{2^{d+1}\pi^{d/2}} N C_{st}(p) \sum_{s=0}^{N-1} \frac{\cos(ps)}{[L(s)]^{\frac{1}{2}d+1}}, \tag{9.103}$$

where

$$L(s) = \sum_{q=2\pi/N}^{2\pi} C_{st}(q) \left[1 - \cos(qs) f(q)\right]. \tag{9.104}$$

Equation (9.103) is a self-consistent equation for the non-ergodicity function $f(p)$. In the vicinity of the bifurcation point the non-ergodicity function $f(p)$ is small and we can expand the right-hand side of (9.103) with respect to $f(p)$. It is shown in the appendix of [215] that because of orthogonality the zero-order term in this expansion vanishes and we arrive at the so-called F_{12}-model, according to the nomenclature of Götze and Sjörgen [220]. In the present book we omit this demonstration for convenience. The extensive numerical analysis of the full (9.103), which is given in the next section, reveals that the bifurcation of $f(p)$ is continuous or of A-type.

To gain a better insight into the Rouse modes freezing mechanism let us consider first a simplified version. This is a toy model which is based on truncation at the level of the two longest modes, $j = 1$ and $j = 2$. In this case the asymptotic form $NC_{st} \approx p^{-1-2v}$, where $p \approx 2\pi j/N \ll 1$, can be used to calculate the coefficients. As a result the toy model equations for $f(1) \equiv f$ and $f(2) \equiv g$ can be recast in the following forms:

$$\frac{f}{1-f} = \Delta_1 f + \epsilon_1 f g,$$

$$\frac{g}{1-g} = \Delta_2 g + \epsilon_2 f^2, \tag{9.105}$$

where the coefficients are

$$\Delta_1 = \Delta N^{2-vd}, \qquad \epsilon_1 = \frac{\Delta N^{2-vd}}{2^{2v}},$$

$$\Delta_2 = \frac{\Delta N^{2-vd}}{2^{2+4v}}, \qquad \epsilon_2 = 2^{1+2v} \Delta N^{2-vd}. \tag{9.106}$$

It is readily seen that in the vicinity of the critical point, $\Delta_1^{crit} = 1$, the coefficient $\Delta_1 = 1 + \sigma$, where $\sigma \ll 1$, and mode amplitudes have the following forms: $f \approx \sigma f_+$ and $g \approx \sigma^2 g_+$, where f_+ and g_+ are constants. The substitution of these forms into (9.105) leads to the solution

$$f(\sigma) = \sigma,$$

$$g(\sigma) = \sigma^2 \frac{\epsilon_2^{crit}}{1 - \Delta_2^{crit}}, \tag{9.107}$$

where it is important that $\Delta_2^{crit} < 1$.

As a result, the trivial solution $f = g = 0$ bifurcates at the critical point $\Delta_1^{crit} = 1$, so that the f mode is linear and the g mode is quadratic with respect to σ. It is obvious that close to the critical point (i. e. $\sigma \ll 1$) the g mode has no effect on the f mode. However, the g mode bifurcates only as a result of f mode bifurcation. In this respect one can say that the Rouse mode freezing follows the "host–slave" scenario. In Section 9.7 we show that this scenario holds true for the whole numerical solution.

9.7 Numerical analysis

In this section we present the numerical solution of (9.103)–(9.104) in the full range of the Rouse mode index j values and for increasing values of the disorder strength Δ, which here acts as a control parameter. As usual in mode coupling theory, complete information about the static correlator, $C_{st}(p)$, is a necessary prerequisite for studying the non-ergodicity equation. In this respect, for a chain of given length, we have numerically solved the static equation (9.83) for $C_{st}(p)$, in which the virial coefficients and the disorder strength Δ are given. By making use of the fast Fourier algorithm we implement the bisection procedure between two trial profiles of $C_{st}(p)$ until convergence to the final solution is achieved. After that we use $C_{st}(p)$ as a static input for the non-ergodicity equation (9.103) and (9.104). Equations (9.103) and (9.104) are solved simultaneously for chain length $N = 128$ as an example.

9.7.1 Bifurcation diagram

We found that for small values of the disorder strength Δ the only solution of (9.103) turns out to be the trivial one, i. e. $f(p) = 0$. As the disorder strength increases above a critical value Δ_{crit}, we observe that the first and all other modes simultaneously become frozen, i. e. they are characterized by a non-vanishing value of the non-ergodicity function $f(p)$ at the same Δ_c. The resulting bifurcation phase diagram is shown in Fig. 9.10. As may be seen from Fig. 9.10 all modes bifurcate continuously (A-type), but bifurcations of higher modes ($j = 2, 3, \ldots$) are smoother than the first mode bifurcation. This is qualitatively consistent with the result of the toy model analysis from the previous section. Moreover, one can see that the higher is the Rouse mode index the smoother is the bifurcation. This creates some numerical difficulties concerning the precise location of the critical point Δ_{crit} of the higher

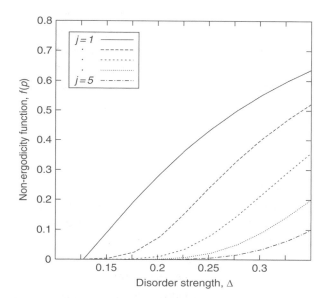

Fig. 9.10. Bifurcation diagram for the non-ergodicity function according to (9.103). The calculation refers to a chain of $N = 128$ monomers and the bare second virial coefficient $v = 0$. We show only the smaller mode index values, i. e. $j = 1, \ldots, 5$. The freezing of the modes appears above a critical value of $\Delta_{crit} \approx 0.13$.

modes. When the accuracy of a non-vanishing value of $f(p)$ close to zero is not sufficiently high, the results in Fig. 9.10 can be interpreted that the different chain modes freeze at different values of the disorder strength (see Fig. 9.11).

The critical value $\Delta_{\text{crit}} \approx 0.13$ should be correlated with the radius-of-gyration diagram in Fig. 9.12. It can be clearly seen that at a disorder strength comparable

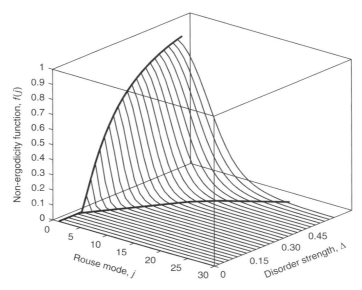

Fig. 9.11. Three-dimensional bifurcation diagram. The illusory Rouse mode successive freezing is a result of the finite resolution, in this case the accuracy of a non-zero value of $f(p)$ or the resolution $h = 10^{-4}$.

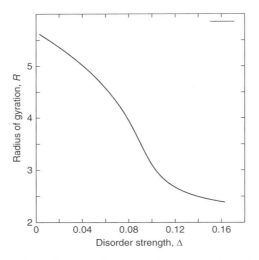

Fig. 9.12. Radius of gyration as a function of disorder strength Δ for the chain length $N = 128$ and $v = 0$.

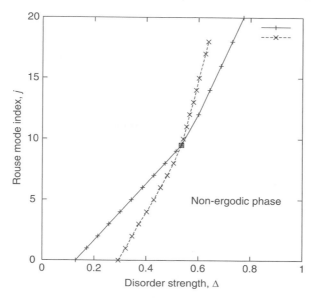

Fig. 9.13. Comparison of the freezing diagram for coiled and globule states. The two curves correspond to conditions with different bare virial coefficients, namely $v = 0$ (solid line) and $v = 0.5$ (dashed line).

with this value the system is approaching the radius of gyration that corresponds to the globular phase [215]. We recall that the static input information which is embraced by (9.103) is determined by the effective virial coefficient $v_{\text{eff}} = v - \Delta$, where v is a bare second virial coefficient. That is why the bifurcation diagram in Fig. 9.10 corresponds to the Rouse mode freezing in the globule phase. It is interesting to explain how the Rouse modes freeze in the coiled state. To do so we have driven the system to the coiled state by increasing the value of the bare virial coefficient to $v = 0.5$ while solving (9.103). The result of the mode freezing is shown in Fig. 9.13 and is compared with the previous case (where $v = 0$). It can be seen that the freezing of the modes in the coiled state at least for small mode indices ($0 < j < 10$) occurs at higher values of Δ.

9.8 Contribution to the modulus

Obviously the dynamics of the localized chains contribute to the elastic and viscoelastic properties of the elastomer. The present theory allows some preliminary predictions. We calculate therefore the contribution of the localized chains to the modulus in the limiting time scales $\tau_1 < t < \tau_1 N^2$. As we stated above, the theory suggests a non-trivial stretched exponential of the form

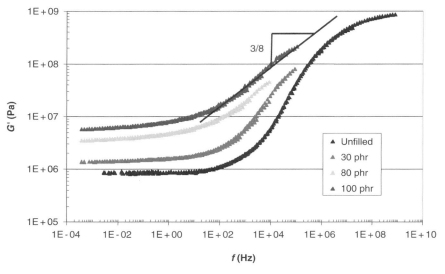

Fig. 9.14. The prediction for the scaling of the frequency dependence of the modulus can be seen for high filler concentrations. There a huge amount of the matrix is localized around and filler clusters and gives $G(\omega) \propto \omega^{3/8}$.

$$G(t) \propto \sum_{q} \exp\left\{-\left(\frac{b^3}{\Delta}\right) \Omega(q) \left(\frac{t}{\tau_1}\right)^{\beta}\right\}, \qquad (9.108)$$

where the value of $\beta = 3/4$ is independent of the disorder. The quantity Ω is given by

$$\Omega \propto \tau_1 q^2. \qquad (9.109)$$

Here we present only a scaling estimate of the modulus. The results for the storage modulus can be summarized as follows:

$$G'(t) \propto \left(\frac{\Delta}{b^3}\right)^{1/2} \left(\frac{t}{\tau_1}\right)^{-3/8}, \qquad (9.110)$$

which transforms into the frequency dependence

$$G'(\omega) \propto \tau_1 \left(\frac{\Delta}{b^3}\right)^{1/2} (\omega\tau_1)^{3/8}. \qquad (9.111)$$

This is the contribution to the modulus from the localized chain dynamics. Note that the disorder contribution to the dynamic modulus shows a very different scaling with respect to the frequency than in the standard Rouse or reptation model (see [62]). In the Rouse model the modulus scales with the frequency as $G_{\text{Rouse}} \propto \omega^{1/2}$. The simple scaling prediction of eq. (9.110, 9.111) agree very well with experimental results, as shown in Fig. 9.14.

10

Filler–filler interaction

10.1 Filler networking in elastomers

10.1.1 Flocculation of fillers during heat treatment

For a deeper understanding of filler networking in elastomers it is useful to monitor structural relaxation phenomena during heat treatment (annealing) of the uncrosslinked composites. This can be achieved by investigations of the time development of the small-strain storage modulus G_0' that provides information about the flocculation dynamics [138, 221–224]. Figure 10.1(a) shows the time development of the small-strain storage modulus G_0' at 0.28% strain and 1 Hz of three elastomer composites containing 50 phr carbon black of different grades. The sample with the smallest primary aggregate size (N115) exhibits the most pronounced increase of the storage modulus with annealing time, which levels out after about 10 minutes in this example. The extent of modulus gain reduces with increasing primary aggregate size and the N550 sample shows almost no effect. With increasing dynamic strain amplitude, as depicted in Fig. 10.1(b), the storage modulus decreases by about one order of magnitude (the Payne effect). Thus, it appears that during heat treatment a weakly bonded superstructure develops in the systems which stiffens the polymer matrix, indicating that the increase of the modulus results from flocculation of primary aggregates to form secondary aggregates (clusters) and finally a filler network. The dependence of the effect on the primary aggregate size is in accordance with the picture of a kinetic aggregation process.

Figure 10.2(a) shows the time development of G_0' of S-SBR melts of variable molar mass filled with 50 phr carbon black (N234), when a step-like increase of the temperature from room temperature to 160 °C is applied. Figure 10.2(b) shows a strain sweep of the same systems after 60 minutes annealing time. Depending on the molar mass M_w, as indicated, a pronounced increase of G_0' is observed in the first minutes that levels out almost to a plateau value at longer annealing times. In agreement with the studies of Wang *et al.* [224], the largest plateau value is observed

153

Filler–filler interaction

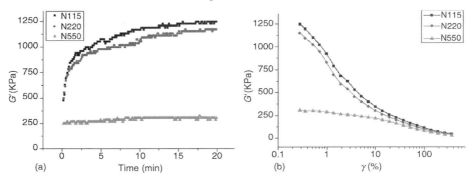

Fig. 10.1. Time development of (a) the small-strain storage modulus (0.28% strain, 1 Hz) of S-SBR (VSL 2525) composites without curatives with 50 phr carbon black during heat treatment at 160 °C for various grades, as indicated, and (b) the strain dependence of the storage modulus of the same samples after heat treatment for 20 minutes.

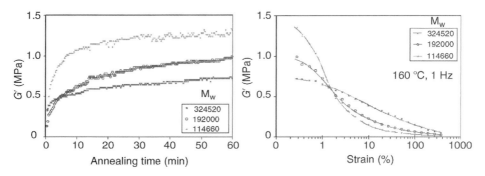

Fig. 10.2. Time development of (a) the small-strain storage modulus (0.28% strain, 1 Hz) of uncrosslinked S-SBR composites with 50 phr N234 during heat treatment at 160 °C for various molar masses, as indicated, and (b) the strain dependence of the storage modulus of the same samples after heat treatment for 60 minutes. From [138].

for the lowest molar mass, confirming that the increase of the modulus results from flocculation of primary aggregates to form secondary aggregates (clusters). With increasing dynamic strain amplitude, as depicted in Fig. 10.2(b), a stress-induced breakdown of the filler clusters takes place and the storage modulus decreases by about one order of magnitude (the Payne effect). With respect to the variable molar mass of the systems, Fig. 10.2 shows a cross-over of the moduli with increasing strain, indicating that a larger molar mass stabilizes the filler–filler bonds more effectively. This can be related to the overlapping action of tightly bound polymer chains in the contact area between adjacent filler particles.

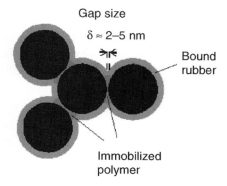

Fig. 10.3. Schematic presentation of a flocculated carbon black cluster in a polymer matrix with characteristic gaps between adjacent particles. The impact of gap size on the stiffness of filler–filler bonds is apparent. The black disks symbolize primary carbon black aggregates.

Based on this kind of flocculation studies and additional dielectric investigations a model of the structure–property relationships of filler–filler bonds in a bulk rubber matrix has been developed [138, 225]. The basic features are illustrated in Fig. 10.3. According to this model, the stiffness of filler–filler bonds is governed by the remaining gap size between contacting particles. This, in turn, depends on the ability to squeeze out the bound polymer chains from the contact area under the attractive action of the depletion force between the filler particles. This process leads to a stiffening of filler–filler bonds. This is favored by several factors, e.g. a high ambient temperature, low molar mass, small particle size, weak polymer–filler and strong filler–filler interaction.

The mechanical connectivity between the filler particles is provided by a flexible nanoscopic bridge of glassy polymer, resulting from the immobilization of the rubber chains in the confining geometry close to the gap. Since the stiffness of the bonds transfers to the stiffness of the whole filler network, the small strain elastic modulus of highly filled composites is expected to reflect the specific properties of the filler–filler bonds. In particular, the small-strain modulus increases with decreasing gap size during heat treatment as observed in Figs. 10.1(a) and 10.2(a). Furthermore, it exhibits the same temperature dependence as that of the bonds, i. e. the characteristic Arrhenius behavior typical of glassy polymers.

In the case of carbon-black-filled diene–rubber composites the polymer–filler interaction is generally quite strong due to the high affinity between the π-electrons at the carbon surface and those in the double bonds of the chains. According to the site energy distribution function estimated in Section 7.2, the typical interaction energy between carbon black and ethene, representing a single double bond, lies between 10 and 35 kJ/mol and depends on the grade number. A more practical procedure for characterizing the polymer–filler interaction in elastomer composites is the estimation of bound rubber i. e. the amount of polymer tightly bound to the filler surface after mixing [226]. It is well known that this amount increases with the molar mass of the polymer and the specific surface area of the filler particles, but is also affected significantly by the surface activity, given, e.g., by the site energy distribution function of the filler obtained with polymer analogous gases [226–229]. A further effect comes from the preparation conditions of the composites, e.g. mixing time [230], since the formation of bound rubber is a slow dynamical process that requires time [138].

10.1.2 Kinetics of filler structures under dynamic excitation

Despite the technological significance of the Payne effect in rubber applications, this strain-induced softening phenomenon is often regarded as a special area of physics specific to filled elastomers. However, dynamic strain-induced non-linearity in the modulus of filled rubbers shows a striking similarity to what is known about the glass transition of solid materials and the jamming transition of granular materials. This analogy stems from the fact that shear strain in dynamic mechanical measurements introduces fluctuations in a filler network by forcing the system to explore different configurations. Such fluctuations can be described by an "effective temperature" that has many of the attributes of a true temperature; in particular it is proportional to the strain amplitude [231,232]. Thus, with respect to strain filled rubbers will display many unusual phenomena that are usually observed in glass-forming materials, including asymmetric kinetics, cross-over effects, and glass-like kinetic transitions.

The asymmetric kinetics in filled rubbers is displayed in Fig. 10.4. The figure shows that the modulus of the filled rubber after abruptly increasing to a certain strain amplitude approaches the steady state faster than that of a rubber released from the higher strain to the lower strain amplitude. Figures 10.5 and 10.6

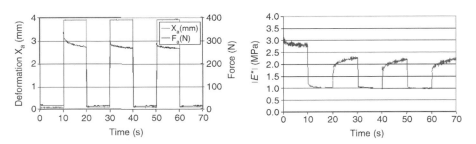

Fig. 10.4. Stiffness response of a filled rubber (50 phr carbon black) on repeated amplitude steps. From [233].

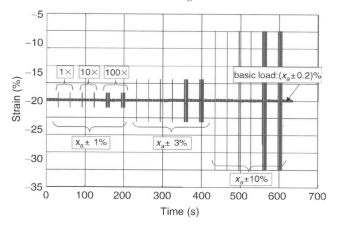

Fig. 10.5. Dynamic-mechanical test program with systematic changes of the strain amplitude during sinusoidal excitations of rubbers in the unidirectional compression mode. Measurements were performed at room temperatures. The periods of basic loading between changes of the amplitude are 30 s; the corresponding loading conditions are 30 Hz, 20% preload and ±0.2% dynamical deformation. The maximal loading increases up to 10%. The frequency during the corresponding maximal loading states was 10 Hz.

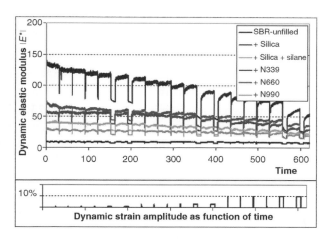

Fig. 10.6. Response of the dynamic elastic modulus according the test program in Fig. 10.5. For unfilled SBR and 50 phr filled samples containing fillers of different reinforcing activity (carbon blacks N990, N660, N339 according to ASTM nomenclature and precipitated silica with and without coupling agent).

show how different kinds of fillers in the same rubber matrix (a styrene-butadiene copolymer (SBR) emulsion) display different "fingerprints" as a reaction against a dynamic–mechanical test program with abrupt changes of the strain amplitude. Such "fingerprints" have importance in the laboratory-based evaluation of filled rubber materials in the rubber and tire industry.

Figure 10.4 shows the stiffness response on repeated amplitude steps (frequency 10 Hz, room temperature). In a very simple and naive approach we may assume that the rate of contacts (say N) contributing to filler–filler interactions after the sudden breakdown due to a step-like increase ($\Delta\varepsilon$) of deformation amplitude is proportional to the square of the concentration of the units. These units correspond to filler aggregates forming successively new agglomerates, respectively the (infinite) filler network [233]. Then,

$$\frac{\mathrm{d}N}{\mathrm{d}t} = kN^2 , \tag{10.1}$$

where k is a rate constant. This type of growth is described as a second-order reaction of type I. We note that it describes not the complete asymmetric behavior of the modulus but the recovery process. The final state would be reached as $N = N_\infty$ at $t \to \infty$. We assume that the corresponding excess elastic modulus $E \equiv |E^*| \ (\approx E')$ (or similarly the shear modulus $G \equiv |G^*| \ (\approx G')$) is proportional to the increase of new contacts, where for $t = t_0 : N = N_0$ and for $t \to \infty :$ $N = N_\infty$. Here, N_0 is the residual number of filler–filler contacts just after the breakdown where the amplitude increases stepwise from ε_0 to $\varepsilon_0 + \Delta\varepsilon$. Therefore, this number N_0 can be regarded as the starting point of the recovery process. The number of contacts after total recovery is N_∞, corresponding to a small (nearly zero) deformation amplitude.

We note that describing the recovery effects by a second-order kinetics in rubber systems is not unusual. The filler–filler interactions have been attributed to the formation of "rubber-like junctions" as depicted in Fig. 10.3 with a glassy layer between two neighboring aggregates forming larger clusters [234]. This layer formation is closely connected with the formation of bound rubber on filler particles. The formation of bound (polydimethyl siloxane, PDMS) rubber on silica has been investigated and a second-order kinetic process was found [235].

Experimentally, after a certain time $\Delta t^* = t_1 - t_0$ the deformation amplitude again increases step-wise from ε_0 to $\varepsilon_0 + \Delta\varepsilon$. As a consequence, the value N_∞ will not be reached. According to the assumptions of a second-order reaction type (equation (10.1)) and the direct relation between the number of contacts and the modulus contribution, the following equation for $E \equiv |E^*|$ immediately follows:

$$\frac{\mathrm{d}E\,(\xi)}{\mathrm{d}\xi} = k\left(\frac{E_\infty - E\,(\xi)}{E_\infty}\right)^2 . \tag{10.2}$$

Equation (10.2) yields the solution for modulus recovery:

$$E\,(\Delta t) = E_\infty\left(1 - \frac{E_\infty}{k\Delta t + C}\right) , \tag{10.3}$$

where C (Pa) is a constant of integration. The rate constant k (Pa/s) depends on the filler type, temperature and filler loading. Equation (10.1) is the "power saturation" case of the hyperlogistic differential equation which is of relevance for aggregate growth in soft systems [236]. Within this very general background, the exponent 2 in (10.1) characterizes the autonomous (but forced by restrictions) saturation in the aggregated growth process and plays the role of complexity measures. The more this exponent deviates from 1, the higher is the cooperativity within the considered system.

In the literature, there are two other systems that are known to show the same asymmetric behavior as their thermodynamic properties approach equilibrium. One is the structural relaxation process in glassy materials, most notably the relaxation and recovery of enthalpy and volume [237, 238]. In that case, the specific volume of a glass after abrupt cooling to a temperature T is known to approach equilibrium faster than that of a glass heated to the same temperature [237]. The reason is that the cooled sample arrives at temperature T with a larger free volume than the heated glass. Another situation with asymmetric relaxation behavior is density fluctuation in a vibrated granular material [239–244]. In this case, the granular material is in its jamming state. Once a void large enough to contain a grain is created, it will quickly be filled by a new particle. The rate of density settling from above or below equilibrium [239] depends on the rate of void creation and the initial density of the material. Nevertheless, in both cases, the free volume plays a crucial rule in determining the rate at which equilibrium is approached. Obviously, these two cases share a common physics ground [242–244].

As already noted, more complicated features are observed in dynamical loading programs like that shown in Figs. 10.5 and 10.6. Such experiments also prove the so-called cross-over effect. Wang and Robertson [231] demonstrated this effect with the following rubber testing program: a highly filled rubber was loaded with a dynamic strain amplitude, say $\gamma_0 = 1\%$, for a period of time t_1 insufficient to reach equilibrium, and then the load was changed to another amplitude, say $\gamma_1 = 7\%$, and the system was allowed to equilibrate. They showed that the storage modulus initially decreases with time and crosses over the actual equilibrium, leading to a surprising minimum that depends upon the prior history of loading applied to the sample. After that, the storage shear modulus G' slowly approaches the equilibrium value. This phenomenon again has only been observed in glass-forming materials in the glassy state [237, 238], and the behavior is usually referred to as a cross-over effect. Loading experiments with rubbers show that granular materials (e.g. fillers) may also display such a cross-over effect if they are impregnated in a soft elastomeric matrix.

The existence of remarkable similarity between dynamic strain-induced non-linearity in the modulus of filled rubbers, the physics of the glass transition of

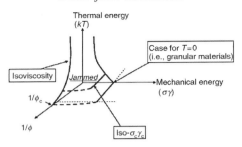

Fig. 10.7. A schematic drawing of the jamming phase diagram. From [232].

glass-forming materials, and the jamming transition of vibrated granular materials
has important implications with regard to our understanding of the strain-induced
non-linearity of filled rubbers. The similarity stems from the fact that filler particles
in the rubber matrix agglomerate and tend to form filler networks. The agglomeration and network formation of the filler in an elastomeric matrix are typically
jamming processes that are analogous to glass formation. It is reasonable that different routes, via strain, volume fraction, and temperature changes, can effectively
lead filled rubbers to the same jammed state. A large unified physical picture describing the dynamics in the frustrated systems would be a jamming phase diagram that
is able to address the glass and jamming transitions [231,232]. Based on the experimental observations, a unified diagram was proposed and is shown schematically in
Fig. 10.7 [231,232]. The phase diagram for isoviscosity lies in the vertical $(1/\Phi)-T$
plane. The line that separates the jammed solids and unjammed liquids generally
represents the glass transition. The classical empirical equation proposed by Doolittle and Doolittle [245] describes approximately the location of this transition. The
transition line marks a critical viscosity of the system that in practice is impossible
to track in the time scales accessible to experiment. The phase diagram for the isoenergetic state lies in the horizontal $(1/\Phi)-\sigma\gamma$ plane. The isoenergetic behavior
comes from experimental observation of various particle-filled systems [232]. The
effects of the temperature on the jamming transition are illustrated in the $T-\sigma\gamma$
plane. The magnitude of the energy is a function of the elasticity of the jammed
fractal structure and the interaction strength between filler particles. The experimental data in [232] show that $\sigma_{crit}\gamma_{crit}$ increases as temperature decreases. The
transition line marks a critical mechanical energy needed to dejam the system.

   Some years ago several authors [242–244] also proposed a phase diagram for
jamming. In their phase diagram, however, other axes were selected, i. e. the temperature T, density Φ, and the shear stress σ; T and Φ are traditional axes for
phase diagrams, but σ is not. The physical origin of selecting σ for the phase diagram is not clear, though mode-coupling theories have attempted to include the
shear stress [246]. In the studies of Wang and Robertson [231,232] it was found

experimentally that using $\sigma\gamma$ instead of σ as the critical parameter can significantly simplify the phase diagram. It is also noteworthy that this mechanical energy $\sigma\gamma$, like the thermal energy kT, is theoretically derivable from Hamiltonians, which makes it a natural choice for an axis when constructing phase diagrams.

To summarize, we have discussed how the non-linearity in the modulus of filled rubbers simply reflects a jamming–unjamming process for fillers in rubber matrices. The agglomeration of filler in an elastomeric matrix shares a common basis in physics with the jamming process and glass formation. Several implications for the behavior of filled rubbers under complex dynamical service conditions (e.g. for tire tread compounds, motor mounts, etc.) can now be better understood. For example, it was found experimentally that aging at a fixed oscillatory strain produces a "hole" in the unjamming loss modulus (G'') spectrum which is localized near an aging strain [247].

10.2 Dynamic small- and medium-strain modeling – the Payne effect

10.2.1 The Kraus model

The Kraus model [35] provides a fairly good quantification of the Payne effect though it contains some rather arbitrary assumptions. The starting point is the idea that the dynamical breakup of the carbon black network can be described by a deagglomeration and an (re) agglomeration rate. Then it is assumed that

(a) the deagglomeration (breakdown) rate R_b is proportional to the number of remaining contacts N between carbon black aggregates and to some power m of the deformation amplitude,

$$R_b = k_b a^m N , \qquad (10.4)$$

(b) the reagglomeration rate R_a in a similar way is proportional to the number of broken contacts $N_0 - N$ (where N_0 is the total number of contacts within the carbon black network at zero deformation)

$$R_a = k_a a^{-m} (N_0 - N) . \qquad (10.5)$$

Here k_b and k_a are rate constants. At equilibrium $R_a = R_b$, the equations can be solved to give

$$N = \frac{N_0}{1 + \left(\frac{a}{a_c}\right)^{2m}} , \qquad (10.6)$$

where $a_c = (k_a/k_b)^{(1/2m)}$ is for the characteristic ratio of the rate constants.

Both assumptions (a) and (b) can be criticized:

(a) Kraus himself pointed out that the deagglomeration rate seems more likely to depend on the deformation rate than on the amplitude. However, this would lead to a strong

frequency dependence of the Payne effect which is not observed experimentally
[182, 248]. Therefore the assumption (10.4) for the breaking rate is justified only by
the success of the results.

(b) It is not immediately obvious that the rate of reagglomeration should (as a function of
the amplitude) decay with the same power m that characterizes the increase of the deagglomeration rate. Instead, one would expect a more general form $R_a = k_a a^{-n}(N_0 - N)$
with an exponent n possibly different from m. This would lead to

$$N = \frac{N_0}{1 + \left(\dfrac{a}{a'_c}\right)^{m+n}} , \tag{10.7}$$

with $a'_c = (k_a/k_b)^{1/(m+n)}$. However, this does not affect the Kraus model because the
exponent m is only a fit parameter without specification or relation to experimentally
measured quantities.

Continuing, the Kraus model assumes that the excess storage modulus is
determined by the momentarily existing contacts,

$$\frac{G'(a) - G'_\infty}{G'_0 - G'_\infty} = \frac{N}{N_0} = \frac{1}{1 + \left(\dfrac{a}{a_c}\right)^{2m}} , \tag{10.8}$$

while the excess loss modulus is proportional to the deagglomeration rate R_b

$$G''(a) - G''_\infty \sim k_b a^m N \sim \frac{a^m (G'_0 - G'_\infty)}{1 + \left(\dfrac{a}{a_c}\right)^{2m}} . \tag{10.9}$$

This results in a somewhat unclear way from the idea that the dissipation of
energy is coupled with the breakup of the filler aggregates. Since the maximum of
the loss modulus G''_{max} is reached at $a = a_c$, the excess loss modulus can also be
written as

$$\frac{G''(a) - G''_\infty}{G''_{max} - G''_\infty} = \left(\frac{a}{a_c}\right)^m \frac{2}{1 + \left(\dfrac{a}{a_c}\right)^{2m}} . \tag{10.10}$$

Now all information about the deformation process is contained in the parameters
m, a_c, G'_0, G'_∞ and G''_∞. The exponent m characterizes the functional form of the
dynamic moduli G' and G''.

The Kraus model has been successfully applied several times to the empirical
description of the dynamic-mechanical behavior of carbon-black-filled vulcanizates
[249, 250]. The exponent m has been found to be universal: it is to a large extent
independent of temperature, frequency, carbon black content, and filler type, see
Fig. 10.8. A fit to experimental data yields the value $m \approx 0.6$.

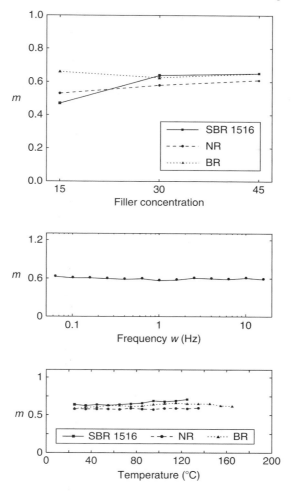

Fig. 10.8. Results for the Kraus model parameter m, from [249]. Unless indicated otherwise, the data refer to butyl rubber filled with 45 phr carbon black N339 at a frequency of 1 Hz and temperature of 25 °C.

The "universality" of filler aggregation, as indicated by the independence of exponent m of concentration, frequency, and a wide range of temperature, is supported further by measurements [251] on elastomeric systems filled with a strongly crosslinked organic material instead of carbon black. Here the Payne effect is found, see Fig. 10.9, and the exponent again has the value $m \approx 0.6$.

Investigations by Ulmer [252] have furnished discrepancies in the loss modulus between experimental results and the expression due to Kraus. The deviations can be eliminated by introducing an additional empirical term, but this will not be

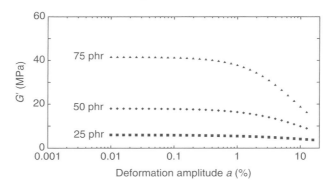

Fig. 10.9. Amplitude dependence of the storage modulus of SBR 1500 with an organic filler (weak coupling of rubber matrix to the filler) at $20\,^{\circ}$C and 1 Hz.

discussed further here, as the present work is mainly concerned with the basic problems of theoretical modeling.

Indeed, the most important shortcoming of the Kraus model is fundamental in nature: the parameters m and a_c are purely empirical, not based on the structure of the filler network. Therefore the Kraus model is not able to provide an explanation of the universal features of the parameters, especially the exponent m.

10.2.2 The viscoelastic model

In the following it is shown that the phenomenological result from the Kraus model [8, 35, 228] can be derived from a physical model that uses realistic assumptions about the filler network structure. This new model is based essentially on the assumption that the clusters forming the filler network have a self-similar, i. e. fractal, structure, which can be described by correlations similar to those in the percolation model. This is not totally correct inasmuch as the cluster growth for carbon black concentrations above the gel point of the filler network is governed by a kinetic cluster-by-cluster aggregation process [181]. Therefore the model presented here is restricted to filler concentrations near the gel point, which is the case for most applications. At lower concentrations a cross-over to pure rubber behavior is expected.

The model introduced in the following enables structure–property relations that reproduce the results of the phenomenological Kraus model to be derived from simple assumptions about the cluster structure. The main idea is that the non-linear behavior of the filled system at dynamical deformation can be described by a viscoelastic model with elements which are non-linear in the deformation amplitude.

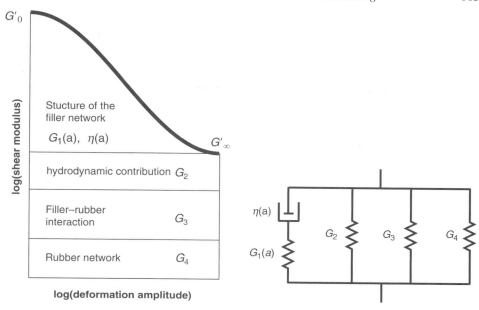

Fig. 10.10. Visualization of how the non-linear behavior of the filled system under dynamical deformation can be described by a viscoelastic model with elements which are non-linear in the deformation amplitude.

To this end we choose as a simple approach an extension of the so-called Zener model [253] which consists of a spring-dashpot (Maxwell) element in parallel with several other springs. Figure 10.10 shows how this model corresponds to the interpretation of the Payne effect mentioned above: the contributions of the rubber matrix, the filler–rubber interaction and the hydrodynamic reinforcement are represented by linear springs, as they can be assumed to be linear elastic at the small deformations that are of interest here; the contribution of the filler agglomerates consists of an elastic part and a viscous part, both of which are non-linear with respect to the amplitude and are taken as a model of the energy storage and dissipation, respectively, during the dynamical breakup of the filler clusters. Frequency ω and temperature T are considered to be constant.

The viscoelastic moduli of the whole system read [253]

$$G'(a) = \sum_{i=2}^{4} G_i + \frac{G_1(a)}{1 + \left(\frac{G_1(a)}{\omega \, \eta(a)}\right)^2} \,, \tag{10.11}$$

$$G''(a) = \frac{G_1(a)^2}{\omega \, \eta(a)} \frac{1}{1 + \left(\frac{G_1(a)}{\omega \, \eta(a)}\right)^2} \,. \tag{10.12}$$

Thus the frame of the viscoelastic model is phenomenological in nature, but the non-linear elements $G_1(a)$ and $\eta(a)$ are determined from the mesoscopic filler structure. They can be expected to have the general form $G_1(a) = G_1 + G_0(a)$ with $G_0(a) = G_0 a^{x_e}$ and $\eta(a) = \eta_0 a^{x_v}$, where G_0, G_1, and η_0 are constants depending on the material properties and the experimental setup (e.g., the disappearance of the Payne effect at low filler concentrations, i. e. the transition to the linear elastic behavior of the rubber matrix, should mainly be governed by the decrease of the constant $G_1 = G_0' - G_\infty'$ at diminishing volume fraction). However, the universal, i. e. material-independent, structure of the filler clusters is reflected in the exponents x_e and x_v. In the following we restrict our considerations to these exponents, i. e. the scaling behavior of $G_1(a)$ and $\eta(a)$.

The basis for the determination of the scaling behavior is the possibility of characterizing the self-similar structure of the filler clusters by means of universal fractal exponents. In principle, we expect the filler agglomerates to show multifractal features because of their statistical nature. But for our purpose characterization by two exponents should be sufficient: apart from the mass fractal dimension d_f, a measure of filler cluster connectivity or degree of branching is necessary. Here we choose the minimum path dimension C which relates the geometrical distance R between points on the cluster with their shortest path length L along the cluster structure [183]:

$$\frac{L}{b} \simeq \left(\frac{R}{b} \right)^C , \tag{10.13}$$

where the filler agglomerates are assumed to consist of smallest units of size b, given by the aggregate size.

As mentioned in Section 10.1, the best description of the carbon black agglomerate structure is obtained with the model of kinetic cluster–cluster aggregation. In this case the fractal exponents are $d_f \approx 1.8$ and $C \approx 1.3$ [254, 261].

Our method is modeled on that of Witten *et al.* [255], who derived a similar scaling relation for the elastic properties of a filled system in the limit of large extensions. These authors started from the idea that the structure of deformed filler clusters can be described by a blob, as known from polymer physics (see e.g. [63]): on length scales above the blob size ξ, the system is homogeneous (the stress is distributed uniformly), on length scales smaller than ξ the stress means only a small perturbation.

This concept can be employed here with only one modification: we assume that the filler clusters are not stretched when subjected to increasing strain but rather are broken up almost immediately into smaller and smaller units, the stress being supported by the rubber matrix. Thus the blob extension can be identified with the (deformation-dependent) cluster size.

In a first attempt to model the Payne effect [179], the cluster size ξ was assumed to be inversely proportional to the external force. This is similar to the well-known assumption used in the original blob model of Pincus in polymer physics (see, e.g., [63]), where the deformation behavior of excluded volume chains is computed. Furthermore, the cluster deformation was neglected totally. In spite of such strong simplifying assumptions this model gives a hint of how the universal exponent m of the Kraus model can be obtained from the filler agglomerate structure. However, the main disadvantage of the model is the lack of an explicit mechanism for the dissipation of energy ($\eta(a)$ is assumed to be constant). Therefore the form of the loss modulus cannot be described correctly.

Here we go beyond this model by taking into account the energetic balance between cluster deformation in the direction of elongation (uniaxial elongation) and perpendicular to it (lateral compression). According to Witten *et al.* [255] the length h of a cluster ensemble created by the breakup of an agglomerate with extension ξ_0 is given by the number $n = N_0/N$ of clusters along the minimum path (where $N_0 \simeq (\xi_0/b)^C$ is the total number of filler aggregates and $N \simeq (\xi/b)^C$ is the number within a cluster) times the cluster size ξ. With (10.13) this means

$$\frac{h}{\xi_0} \simeq \frac{\xi}{\xi_0} \frac{N_0}{N} \simeq \left(\frac{\xi_0}{\xi}\right)^{C-1} .$$

Because the amplitude a is proportional to the deformation, the width w of the cluster ensemble is obtained as $w \simeq \xi_0 a^{-1/2}$ for uniaxial deformation in $d = 3$. The volume occupied by the clusters is $hw^2 \simeq \xi^3(\xi_0/\xi)^{d_f}$. These two expressions together lead to a relation between cluster size and deformation amplitude:

$$\frac{\xi}{\xi_0} \sim a^{-1/(C-d_f+2)} . \tag{10.14}$$

Now the scaling relations for the non-linear elements of the viscoelastic model can easily be determined: the viscosity as a measure of energy dissipation in analogy to the friction coefficient [62, 63] is taken to be proportional to the cluster extension, $\eta \sim \xi/\xi_0$, leading to

$$\eta(a) \simeq \eta_0 \, a^{-1/(C-d_f+2)} . \tag{10.15}$$

The energy U stored in a volume ξ_0^3 results from the energy stored in a cluster of size ξ (which is proportional to $(\xi_0/\xi)^C$) times the total number of clusters $(\xi_0/\xi)^{d_f}$, yielding $U \sim a^{(C+d_f)/(C-d_f+2)}$ with (10.14). For $G_0(a) \sim (1/a)(\partial U/\partial a)$ this finally leads to a scaling relation for the non-linear part of the elastic modulus of the filler:

$$G_0(a) \sim a^{\frac{3d_f-C-4}{C-d_f+2}} . \tag{10.16}$$

Now, if the numerical values of the fractal dimensions $d_f \approx 1.8$ and $C \approx 1.3$ for the case of CCA clusters are inserted, the exponent assumes the value 0.066. This means that the elasticity of the filler clusters is practically linear in the amplitude regime between 0.001% and 10% relevant for the Payne effect. Therefore in the following we set $G_1(a) \approx G_1 + G_0$ for comparison with the Kraus model.

These results for the elastic and viscous parts of the filler network, (10.11), lead to $G'_\infty = \sum_{i=2}^{4} G_i$ and $G'_0 - G'_\infty = G_1 + G_0$. This together with (10.12) produces the excess moduli

$$\frac{G'(a) - G'_\infty}{G'_0 - G'_\infty} = \frac{1}{1 + K^2 a^{2/(C-d_f+2)}}, \tag{10.17}$$

$$\frac{G''(a)}{G'_0 - G'_\infty} = \frac{K\, a^{1/(C-d_f+2)}}{1 + K^2 a^{2/(C-d_f+2)}}, \tag{10.18}$$

where $K = (G_1 + G_0)/(\omega\, \eta_0)$ is a constant containing the system parameters (frequency, temperature, material properties).

Obviously the excess moduli (10.17) and (10.18) derived from the viscoelastic model have the same functional form as the results (10.8) and (10.9) of the Kraus model, if the form exponent m is written as

$$m = \frac{1}{C - d_f + 2}. \tag{10.19}$$

Thus, after insertion of the numerical values for CCA clusters, $d_f \approx 1.8$ and $C \approx 1.3$, the form exponent is calculated as $m \approx 0.66$, compared to the experimental value $m_{exp} \approx 0.6$. The slight discrepancy can be explained by the approximations made within the viscoelastic model: first, the clusters are assumed as monodisperse; second, the CCA model is only a first approach to the true structure of filler agglomerates.

The Zener model applied here should be taken as an idealized approach, chosen with regard to our main point of interest, i. e. the modeling of the amplitude dependence. Therefore we do not expect the model to reproduce other features of filled systems in a similarly realistic way. In the first instance the model parameter η_0 cannot clearly be mapped on experimentally found quantities, since it is itself expected to depend on the frequency ω of periodic deformation – otherwise the frequency dependence of the elastic moduli, which is known to be weak experimentally [182, 248], would be overestimated by (10.17) and (10.18). In order to gain a full picture, realistic forms for the rubber behavior and the filler–rubber interaction have to be included in the viscoelastic model instead of the Hookean springs.

Nevertheless, with the viscoelastic model in its present simple form we have been able to derive the characteristic – hitherto empiric – form of the amplitude dependence of the dynamic elastic moduli from the structure of the filler network. Thus the model confirms the universality of filler aggregation to be responsible for the universality of the exponent m (also in the case of fillers other than carbon black, see Fig. 10.9), and in this way fills a gap in our understanding of the Payne effect.

The main advantage of this model is that it allows the characterization of strain sweep experiments, in the sense that the exponent used in the Kraus model does not depend on the nature of the filler particles. Hence we expect the same behavior for all types of filler particles, independent of their special surface interactions, as long as they form clusters.

10.2.3 The van der Walle–Tricot–Gerspacher (WTG) model

Van der Walle, Tricot and Gerspacher [36] developed a weighted aggregate contact model to describe the low-strain dynamic properties of filled rubbers. In this first the strain dependence of the storage and loss moduli of an elementary two-aggregate system are calculated. Then the complex moduli of the macroscopic system are derived by introducing a weighting function $W(a)$, which determines the relation between $G'(a)$ and $G''(a)$ (a being the strain amplitude).

The WTG model is restricted to the investigation of the non-linear behavior of the interaggregate contacts. The force between interaggregate contacts is assumed to be of the London–van der Waals type. It is not the precise form of the force which is important for the model, but the existence of two stable equilibrated states for a pair of neighboring aggregates, i. e. a bound and an unbound state, respectively. Then the decrease of the storage modulus at cyclic deformation can be explained by the breaking of aggregate contacts, which is equivalent to the transition from the bound to the unbound state, whereas the maximum of the loss modulus results from friction effects/dissipation/slippage in the polymer matrix during the transition between the stable states.

The complex modulus g^* of an idealized model, which contains only two point-like aggregates, is calculated from the hysteresis cycle of the force $F(a)$ between aggregates as a function of amplitude. Let a_b be the deformation amplitude above which the contact between the two aggregates breaks. For $a < a_b$, there is no energy loss since the hysteresis has not yet taken place. At the onset of the hysteresis cycle (at $a = a_b$) the energy loss, i. e. the area enclosed by the hysteresis cycle, can be approximated by $E_l \sim g_0 a_b \, \Delta a$, where g_0 is the slope of the increasing part of force $F(a)$ and Δa is the width of the hysteresis cycle.

In a linear elastic system the energy loss, averaged over a deformation cycle, is connected with the loss modulus through the relation $G'' \sim a^{-2}E_1$ [253]. Using this analogy, van der Walle *et al.* obtained the loss modulus of the elementary pair of aggregates,

$$g''(a, a_b) = \begin{cases} 0 & \text{for } a \leq a_b \\ g_0 \dfrac{\Delta a}{a_b} \dfrac{a_b^2}{a^2} & \text{for } a > a_b \end{cases}.$$

Estimation of the storage modulus by linear regression of $F(a)$ on the interval $[0, a]$ gives

$$g'(a, a_b) = \begin{cases} g_0 & \text{for } a \leq a_b \\ g_0 a_b^3/a^3 & \text{for } a > a_b \end{cases}.$$

Thus the microscopic viscoelastic function $g^*(a, a_b) = g'(a, a_b) + ig''(a, a_b)$ for the idealized model is determined as a function of the amplitude, but nothing is known about g_0 and Δa.

Now in the composite there is a distribution of the orientation and separation distances of the aggregate contacts with respect to the direction of the applied strain. As a result, the composite can be considered as a collection of many elementary models, each having different a_b. This distribution of breaking amplitudes is taken account of by introducing a function $N(a_b)da_b$, which gives the number of links that break when the amplitude is increased from a_b to $a_b + da_b$. The two parameters g_0 and Δa also depend on a_b since they differ for various aggregate contacts. As the effect of a large number of weak links is indistinguishable from that of a small number of strong links, it is convenient to combine $g_0(a_b)$ and $N(a_b)$ into a weighting factor $W(a_b) = g_0(a_b)N(a_b)$. The dynamic moduli are then given by

$$G'(a) = G'_\infty + \int\limits_a^\infty da_b\, W(a_b) + \int\limits_0^a da_b\, \frac{a_b^3}{a^3} W(a_b), \tag{10.20}$$

$$G''(a) = G''_\infty + \frac{h}{a^2} \int\limits_0^a da_b\, a_b^2\, W(a_b), \tag{10.21}$$

where the ratio $h = \Delta a/a_b$ of the width of hysteresis and the breaking amplitude is taken to be constant. $G^*_\infty = G'_\infty + iG''_\infty$ designates the constant contribution of the elastomer matrix.

The fit to experimental data takes place as follows: from experimental curves of $G''(a)$ the weighting function $W(a_b)$ and the constant h are determined. Then a theoretical form $G'_{th}(a)$ of the storage modulus can be calculated, which is to be compared with the measured values $G'_{exp}(a)$. The agreement between calculated and measured values of the storage modulus turns out to be fairly good for all types of carbon black. Thus the WTG model very clearly confirms the interdependence of storage and loss modulus at small deformations. However, the WTG model provides no insight into the relation between filler structure and dynamic mechanical properties, as only the elementary pair of aggregates shows universal and material-independent features. All information concerning the filler agglomeration and other material properties is contained in the fit parameters $W(a_b)$ and h, which are purely empirical. The weighting function can be qualitatively explained as a measure of the number of remaining contacts during cyclic deformation (it decreases with growing amplitude as the number of broken contacts increases), but a more precise form of this relation is not available so far.

10.2.4 The links–nodes–blobs (LNB) model

The model developed by Lin and Lee [37, 268] leads to expressions for the moduli that are quite similar to those from the WTG model, surpasses it by establishing the connection between the involved weighting function and the elasticity of the percolating filler network. The starting point is the assumption that the filler network can be described in terms of the links–nodes–blobs (LNB) model, which is well known from percolation theory [256]:

- The blobs correspond to dense filler clusters which are rigid enough not to be deformed throughout the strain cycle. A blob therefore can be a primary aggregate (this is the smallest possible size) or a cluster formed by coagulation of primary aggregates, occluded rubber, and bound rubber.
- The links correspond to tenuous filler bonds between dense filler aggregates (blobs), consisting of singly connected bonds in terms of percolation theory. They are deformed under tension, bending and/or torsion, and even tend to break off when some failure strain is applied. The smallest link corresponds to a direct contact bonding between two dense filler aggregates.
- Blobs and links together build up so-called LNB chains, in which the connecting points are called nodes. The average length ξ_p of an LNB chain between two nearest nodes corresponds to the critical length of percolation. Therefore, the system is uniform on macroscopic length scales $> \xi_p$.

The basic assumption of the model is that a blob does not deform and an LNB chain can only break at singly connected bonds on a link.

The derivation of the elastic moduli is based on the two-dimensional model by Kantor and Webman [257] for the elasticity of an LNB chain containing L_1 singly connected bonds (see Section 10.3.2). In realistic systems this number is distributed according to a (density) distribution function $f_1(L_1)$. After extending the Kantor–Webman model to three dimensions, Lin and Lee obtained an expression for the (static) modulus of the LNB chain that depends on $f_1(L_1)$, and another expression for the failure strain amplitude (for breaking direct contacts) that depends on L_1.

In the next step different assumptions were made concerning the sequence of breaking and recombination during a deformation cycle. The best result was obtained with the (not very realistic) assumption that half of the recombinations occur when the deformation reaches its maximum and the other half when it reaches its minimum. In this case the dynamic moduli read

$$G'(a) = G'_\infty + K \int_{2a}^{\infty} da_1 \, \frac{f_1(a_1)}{a_1} \,, \tag{10.22}$$

$$G''(a) = G''_\infty + \frac{K}{2\pi a^2} \int_{2a_{\mathrm{app}}}^{2a} da_1 \, a_1 \, f_1(a_1) + (G''_0 - G''_\infty) \int_{2a}^{\infty} da_1 \, f_1(a_1) \,, \tag{10.23}$$

where K and a_{app} are parameters that depend on the local elastic constants, the bond length, the average number of singly connected bonds between two blobs, and ξ_p. Furthermore, $f_1(a_1)$ is related to $f_1(L_1)$ through the variable transform $L_1 = a_1/a_{\mathrm{app}}$.

The similarity with the results (10.20) and (10.21) of the WTG model comes into clear focus if we set $W(a_b) = K f_1(2a_b)/a_b$; the differences can be explained as an effect of the different assumptions concerning the breaking and recombination sequence.

The advantage of the model is that, in principle, $f_1(L_1)$ contains all the information on the structure of the percolation clusters. Nevertheless the precise form of $f_1(L_1)$ is again not available. In particular, the dependence on the universal properties of percolation is not clear. This turns out to be a major disadvantage. Lin and Lee assumed an exponentially decaying distribution of $f_1(L_1)$ which contains only one free parameter. But this arbitrary choice yields a fit which is significantly less successful than the results of the Kraus model or the WTG model. Though the local parameters of the model can be determined from the fit, the interpretation of the results is difficult: on the one hand, the mapping of the concept of links and singly connected bonds on structures in real filler systems is problematic; on the other hand, the LNB model for the description of percolation clusters is valid only for space dimensions $d \geq 6$ [256].

10.2.5 The model of the variable network density

In contrast to all other models, in the model of the variable network density of Maier and Göritz [258] the Payne effect is not related to the breaking and recombination of the filler network. Instead, the behavior of the moduli is attributed exclusively to the assumption of different binding strengths for the polymer chains at the filler surfaces. Thus the main effect of the filler particles should be an increase of the network density. For dynamical deformation, the network density decreases with growing deformation amplitude, as more and more of the weakly bound (i. e. unstable) chains are torn off the filler surfaces. This mechanism is based on the idea that chain adsorption takes place at a specific number of interaction sites per surface unit, with the sites gradually being occupied by the polymer chains. Hence the chains that "arrive first" should have a larger number of possible adsorption sites, thus be bound more strongly than "later" chains which find the surface area widely covered.

Accordingly the storage modulus as a function of deformation amplitude a reads

$$G'(a) = (N_c + N_{st} + N_i(a))k_B T ,$$ (10.24)

where N_c denotes the chemical network density, N_{st} the network density caused by stable bonds between chains and filler and $N_i(a) = N_{i_0} f(a)$ the amplitude-dependent contribution of weakly bound chains.

For the calculation of the unstable contribution, the mechanism of adsorption and desorption of the chains is assumed to be analogous to Langmuir adsorption. Then the procedure is formally quite similar to the Kraus model: instead of breaking and recombination rates, now adsorption and desorption rates are brought into balance. Additionally, the desorption rate is assumed to be linearly proportional to the amplitude, whereas the adsorption rate is assumed to be constant, so that a constant value 0.5 appears instead of the exponent m of the Kraus model,

$$\frac{G'(a) - G'_\infty}{G'_0 - G'_\infty} = \frac{1}{1 + ca} .$$ (10.25)

Here, similarly to the Kraus model, the parameter c, is a purely empirical rate constant (more precisely it is the ratio of the constant part of the desorption rate to the adsorption rate). $G'_\infty = (N_c + N_{st})k_B T$ is interpreted as the constant contribution from stable bonds and crosslinks and $G'_0 - G'_\infty = N_{i_0}k_B T$ as the contribution from unstable bonds.

Maier and Göritz assumed the loss modulus to be proportional to the product of the number of occupied and free interaction sites, leading to

$$G''(a) = G''_\infty + G''_i \frac{ca}{(1 + ca)^2} ,$$ (10.26)

where G''_∞ and G''_i are the contributions of stable and unstable bonds, respectively. In this case the deviation from the Kraus model is somewhat more evident (compare (10.10)), but the quality of fits to the experimental data shows no noticeable difference.

As energy dissipation at deformation is always accompanied by spatial reorganization and slippage of the chains, the idea of there being varying bond strengths might be an interesting approach. This holds even if the dynamical picture of chains that adsorb at different times is abandoned (as yet little is known about the adsorption process on mixing the filler into the polymer melt). But a model that adopts this mechanism for the only origin of the Payne effect, totally neglecting the existence of a filler network, has to be rejected. First, it is difficult to understand why the deformation amplitudes up to ca. 10% which are relevant for the Payne effect should be sufficient to tear off all the weakly bound chains from the filler surfaces – this concept is not compatible with the existence of a bound rubber layer. Second, the model cannot explain the more than linear increase of the modulus G'_0 with growing filler content, which is an effect of the percolating filler network, but which here is only interpreted as the contribution of the unstable bonds. Third, the model cannot give a convincing explanation for the existence of the Payne effect in silica-filled rubbers activated with silane (see Section 7.3) and in carbon-black-filled non-polymeric systems such as fluid paraffin and n-decan [248].

10.2.6 The cluster–cluster aggregation (CCA) model

CCA of filler particles in elastomers takes into account that the particles in a rubber matrix are allowed to fluctuate around their mean position. The fluctuation length compares with the rubber-specific fluctuation length of the chain segments, i. e. the mean spacing of successive chain entanglements. Upon contact with neighboring particles or clusters they stick together, irreversibly, since the thermal energy of colloidal particles is small compared with the interaction energy. Depending on the concentration of filler particles, this aggregation leads to spatially separated clusters or a filler network that can considered to be a space-filling configuration of fractal CCA clusters. The two cases are shown schematically in Fig. 10.11(b) and (c).

At low filler concentrations ϕ, below the gel point ($\phi < \phi^*$), the cluster growth with increasing filler concentration is assumed to follow a power law with respect to a net concentration $\phi - \phi^+$. In terms of the solid fraction ϕ_A of the clusters, the power-law dependence reads [151]:

$$\phi_A(\phi) = \phi_P \left(1 + \beta'\left(\phi - \phi^+\right)^B\right)^{-1} \quad \text{for } \phi^+ < \phi < \phi^*. \quad (10.27)$$

Here, ϕ^+ is the critical concentration at which aggregation starts, β' is a scaling factor, and B is a rubber-specific growth exponent. The aggregation limit ϕ^+ is

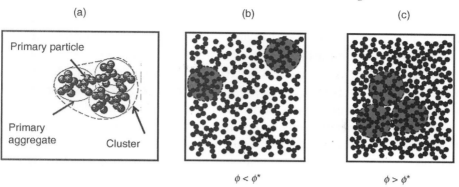

Fig. 10.11. Schematic view of kinetically aggregated filler clusters in rubber below and above the (mechanical) gel point ϕ^*. (a) The local structure of carbon black clusters, built by primary particles and strongly bonded primary aggregates. Accordingly, every disk in (b) and (c) represents a primary aggregate (from [138]).

expected to depend on the fluctuation length of the particles as well as on the compatibility of the rubber and the particles. In addition, it is strongly influenced by the particle size (diameter d) via a critical mean particle distance d^+ as follows:

$$\phi^+ = \alpha \left(\frac{d}{d^+ + d} \right)^3 . \tag{10.28}$$

The value of the factor α depends on the arrangement of particles, e.g. $\alpha = \pi/6$ in a simple cubic lattice. Obviously, the cluster growth for $\phi < \phi^*$ is governed by the restricted mobility of filler particles in the rubber matrix.

At high filler concentrations, above the gel point ($\phi > \phi^*$), the filler particles come sufficiently close together that there is no effect of restricted particle mobility on the cluster growth. Under this condition a kinetic cluster-by-cluster aggregation leads to a space-filling configuration of CCA clusters, similar to colloid aggregation in low-viscosity media [133, 142, 255, 259]. Due to the characteristic self-similar structure of the CCA clusters with fractal dimension $d_f \approx 1.78$, the cluster growth as described by the solid fraction ϕ_A of the clusters is given by a space-filling condition, stating that the local solid fraction equals the overall solid filler concentration:

$$\phi_A (\xi) \cong \phi \qquad \text{for } \phi > \phi^* . \tag{10.29}$$

The solid fraction of the fractal CCA clusters obeys the scaling law:

$$\phi_A (\xi) = \frac{N (\xi) d^3}{\xi^3} \cong \left(\frac{d}{\xi} \right)^{3 - d_f} . \tag{10.30}$$

Here, N is the number of particles of size d in the clusters of size ξ. Equations (10.29) and (10.30) imply that the cluster size ξ decreases with increasing filler concentration ϕ according to a power law. This reflects the fact that smaller clusters occupy less empty space than larger ones (the space-filling condition). It means that the size of the fractal heterogeneities of the filler network, shown as dashed circles in Fig. 10.11(c), decreases with increasing filler concentration.

For low filler concentrations, below the gel point ϕ^*, where isolated particles ($\phi < \phi^+$) or clusters ($\phi > \phi^+$) are considered (Fig. 10.11(b)), reinforcement of rubber is due to hydrodynamic amplification by the filler. Accordingly, stress between the particles or clusters is transmitted by the relatively soft rubber matrix leading to an overstraining of the rubber. If the particles or clusters are sufficiently rigid, i. e. their elastic modulus G_A is much larger than that of the rubber G_R ($G_A \gg G_R$), then the large majority of the elastic energy is stored in the rubber and the elastic modulus G of the composite can be approximated by a linear function of G_R [151]:

$$G \approx G_R X \ . \tag{10.31}$$

The hydrodynamic amplification factor X relates the intrinsic strain γ_R of the rubber to the external strain γ of the sample ($X = \gamma_R / \gamma$). It can be expressed by a power law series as follows [46, 151]:

$$X = 1 + 2.5\phi_{\text{eff}} + 14.1\phi_{\text{eff}}^2 + \cdots \ , \tag{10.32}$$

with

$$\phi_{\text{eff}} = \frac{\phi}{\phi_A\,(\phi)} \ . \tag{10.33}$$

The solid fraction ϕ_A of the clusters is given by (10.27), if $\phi^+ < \phi < \phi^*$, while $\phi_A = 1$ for $\phi < \phi^+$. The occluded rubber concept of Medalia [260] can be applied to consideration of the effective volume fraction in (10.32). This concept assumes that part of the rubber in the voids of the clusters is shielded from deformation and acts like additional filler material. This effect is taken into account in (10.33). It must be noted that the rigidity condition $G_A \gg G_R$ is a necessary condition for rubber reinforcement by filler clusters, because a structure that is softer than the rubber cannot contribute to the stiffening of the polymer matrix. We will see in the next section that the rigidity condition is not fulfilled in all cases, because the modulus G_A of the clusters decreases rapidly with increasing size of the clusters. This means that relatively small filler clusters of less than 100 particles can lead to reinforcement of the polymer matrix with $G_R \cong 0.1$ MPa. Throughout this section we will consider only the case that is necessary for reinforcement, i. e. we assume that the rigidity condition $G_A \gg G_R$ is fulfilled.

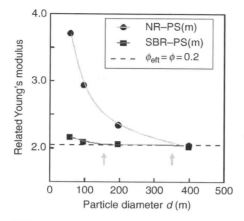

Fig. 10.12. Related Young's modulus of NR and SBR composites with PS-microgel particles (PS(m)) of different sizes at constant volume fraction $\phi = 0.2$ [29].

Much insight into the aggregation of colloidal fillers in elastomers is provided by comparing model fillers with a fairly good filler–rubber interaction with those incorporated in a matrix with a poor filler–rubber interaction, which corresponds to an increased interfacial tension. As shown in Fig. 10.12, the samples obtained by dispersing polystyrene microgels (PS(m)) of different particle size (from 60 to 400 nm) at $\phi = 0.2$ in SBR and NR show an increase in Young's equilibrium modulus at small strain with decreasing particle size above the predicted value of the hydrodynamic theory of Einstein–Smallwood–Guth–Gold (dashed line) [46]. This indicates the formation of rigid microgel clusters below a rubber-specific critical particle size. From Fig. 10.12 it can be concluded that the critical particle diameter below which PS(m) tends to form clusters is 350 nm in NR, and 150 nm in SBR. Thus, with $\alpha = \pi/6$, (10.28) implies a critical mean particle distance of $d^+ \approx 132$ nm for the NR–PS(m) system and $d^+ \approx 57$ nm for the SBR–PS(m) system. The ranking of these values correlates with the δ parameter difference of the corresponding blend constituents, which is proportional to the interfacial tension. In addition, a correlation with the in-rubber particle mobility, as given by the mean distance between successive chain entanglements, d_0, in the two rubbers is obtained if the relation for the plateau modulus $G_N^0 \sim d_0^{-2}$ together with the values $G_N^0 \approx 0.45$ MPa for bulk NR and $G_N^0 \approx 0.9$ MPa for bulk SBR is considered.

Due to the extensive mixing cycle of 30 minutes it can be assumed that in both systems, PS(m)–SBR and PS(m)–NR, the polymeric filler is quite well dispersed. Once the shear forces have been set during mixing, the filler particles aggregate if the interparticle distance is smaller than a critical size d^+. The joining together

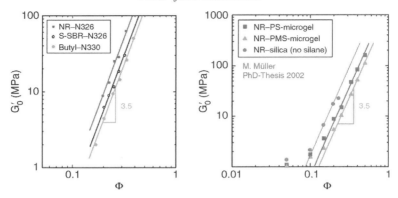

Fig. 10.13. Scaling behavior of the small strain modulus vs. filler volume fraction of (a) carbon black composites and (b) microgel or silica composites. In all cases an exponent close to 3.5 was found, indicating the universal character of the CCA model. From [138].

of a number of microgel particles leads to some extent to the shielding of a part of the rubber matrix from elastic deformation. For that reason, Young's modulus increases due to hydrodynamic amplification as given by (10.31)–(10.33) with an effective volume fraction ϕ_{eff} that varies with particle size d according to (10.27) and (10.28). By inserting $\phi = 0.2, d^+ = 132$ nm and referring to the data presented in Fig. 10.12, the cluster growth function $1 + \beta' \left(\phi - \phi^+ \right)^B$ for the NR–PS(m) system has been estimated. A least-squares fit yields $\beta' = 362$ and $B = 3.7$ with a correlation coefficient close to 1 ($r^2 = 0.982$). This demonstrates the significance of particle aggregation according to the power law (10.27) if a critical concentration ϕ^+ or particle distance d^+ is exceeded.

For filler concentrations above the gel point, $\phi > \phi^*$, where a through-going filler network is formed (Fig. 10.11(c)), stress between the closely packed CCA clusters is transmitted directly between the spanning arms of the clusters that bend substantially. For that reason the strain of the rubber is almost equal to the strain of the spanning arms of the clusters ($\gamma_{\mathrm{R}} \approx \gamma_{\mathrm{A}}$). This means that, due to the rigidity condition $G_{\mathrm{A}} \gg G_{\mathrm{R}}$, the great majority of the elastic energy is now stored in the bent arms of the clusters and the contribution of the rubber to the elastic modulus G of the sample can be neglected, i. e. $G \approx G_{\mathrm{A}}$. This indicates that the stored energy density (per unit strain) of highly filled elastomers can be approximated by that of the filler network, which in turn equals the stored energy density of a single CCA cluster. The last conclusion follows from the homogeneity of the filler network on length scales above the cluster size ξ.

For an estimation of the concentration dependence of the elastic modulus G it is necessary to consider the elastic modulus G_{A} of the CCA clusters more closely.

By referring to the Kantor–Webman model [257] (see Section 10.3.2), we obtain the elastic modulus of the elastically effective CCA cluster backbone from the bending–twisting modulus of tender curved rods [138, 151, 255, 261]:

$$G_A \cong G_P \left(\frac{d}{\xi}\right)^3 N_B (\xi)^{-1} \cong G_P \left(\frac{d}{\xi}\right)^{3+d_{f,B}} \cong G_P (\phi_A)^{(3+d_{f,B})/(3-d_f)}. \quad (10.34)$$

Here, $G_P \cong \overline{G}/d^3$ is the elastic modulus of the cluster units, which is related to the bending–twisting force constant $\overline{G}$ of the bonds between filler particles, N_B is the particle number in the cluster backbone and $d_{f,B} \approx 1.3$ is the fractal dimension of the CCA cluster backbone [142, 259]. The local elastic constant $\overline{G}$ is assumed to be controlled by the stiffness of the polymer bridges between adjacent filler particles, i. e. the properties of filler–filler bonds (see Section 10.1.1).

Equation (10.34) describes the modulus G_A of the clusters as that of its units G_P times a factor that involves the size and geometrical structure of the clusters. The first equality uses the definition of the fractal dimension $d_{f,B}$ $\left(N_B \cong (\xi/d)^{d_{f,B}}\right)$ while the last equality follows from (10.30). If (10.34) is combined with (10.29) we find the following dependence of the elastic modulus G of the composite on filler volume fraction:

$$G \cong G_P (\phi)^{(3+d_{f,B})/(3-d_f)}. \quad (10.35)$$

Equation (10.35) predicts the power-law behavior $G \sim \phi^{3.5}$ for the elastic modulus. Thereby, the exponent $(3 + d_{f,B})/(3 - d_f) \approx 3.5$ reflects the characteristic structure of the fractal heterogeneities of the filler network, i. e. the CCA clusters with $d_f \approx 1.78$ and $d_{f,B} \approx 1.3$ [142].

The predicted scaling behavior, (10.35), of the small-strain modulus with exponent 3.5 is found to be fulfilled for many elastomer composites [123, 138, 151, 261], including the classical butyl–N330 data of Payne [11]. This is demonstrated in Fig. 10.13, in which the small-strain moduli of various filled rubbers at different concentrations are shown as a log–log plot. Obviously, the power-law dependency with exponent 3.5, predicted by the CCA model above the gel point ϕ^*, is found to be fairly well satisfied.

Equation (10.35) represents a scaling relation for the concentration dependence of the elastic modulus of highly filled rubbers that is independent of particle size. This invariance results from the special invariant form of the space-filling condition, (10.29), together with the scaling invariance of (10.30) and (10.34), where the particle size d enters as a normalization factor for the cluster size ξ, only. The scaling invariance disappears if the effect of the immobilized rubber layer that increases linearly with the specific surface of the filler is considered. The immobilized rubber layer can have a pronounced influence on the elastic modulus G_A of the CCA clusters even if the thickness Δ of the layer is small, because the particle size d with the large exponent $3 + d_{\mathrm{f,B}}$ enters into (10.34). The effect of a hard "glassy" layer of immobilized polymer on the elastic modulus of CCA clusters can be modeled, semi-empirically, by introducing a mechanically effective solid fraction $\widetilde{\phi}_A$ of the clusters like in (10.30):

$$\widetilde{\phi}_A (\xi) \approx \frac{N(\xi) \left\{ \frac{\pi}{6} (d + 2\Delta)^3 - \frac{2\pi}{3} \Delta^2 \left(3 \left(\frac{d}{2} + \Delta \right) - \Delta \right) \right\}}{\frac{\pi}{6} \xi^3}$$

$$\approx \frac{(d + 2\Delta)^3 - 6d\Delta^2}{d^3} \phi_A (\xi) \quad \text{for } \Delta \ll d \,. \tag{10.36}$$

This equation considers the mechanically effective solid volume of the clusters, approximately, by enlarging the particle diameter from d to $d + 2\Delta$ and subtracting the volume $V = (2\pi/3)\Delta^2 \left(3 \left(d/2 + \Delta \right) - \Delta \right)$ that results from the intersections of the layers of thickness Δ at the contact points with two neighboring particles. The second approximate equality in (10.36) follows with (10.30) and neglects the high-order Δ terms in the second summand. It is obvious that (10.36) neglects three-fold intersections of particles that are very close or form small loops. Furthermore, the intersections that result from intercluster contacts in a space-filling configuration of CCA clusters are not included.

The mechanical action of the immobilized rubber layer on spherical filler particles, which are assumed to form a filler network in a rubber matrix for $\phi > \phi^*$, is obtained if the mechanically effective solid fraction $\widetilde{\phi}_A$, given by (10.36), is applied in (10.34) instead of ϕ_A and the space-filling condition $\phi_A \cong \phi$ is used. This yields the following power-law dependence of the elastic modulus G on filler concentration ϕ, particle size d, and layer thickness Δ:

$$G \cong G_P \left(\frac{(d + 2\Delta)^3 - 6d\Delta^2}{d^3} \phi \right)^{(3 + d_{\mathrm{f,B}})/(3 - d_{\mathrm{f}})} . \tag{10.37}$$

This equation predicts a strong impact of layer thickness Δ on the elastic modulus G while the influence of particle size increases if d becomes smaller and approaches the value of Δ. Due to the neglect of intercluster contacts in (10.36) and (10.37), the factor in front of ϕ that describes the impact of the immobilized rubber layer is independent of filler concentration. Including intercluster contacts would result in a concentration dependence for this factor, because the number of intercluster contacts increases with increasing filler concentration. This would modify the concentration dependence of the elastic modulus as predicted by (10.37), leading to a somewhat lower value of the elasticity exponent.

At higher filler concentrations, above the gel point ϕ^*, the cluster growth is given by the space-filling condition (10.29) and the main contribution to the elastic

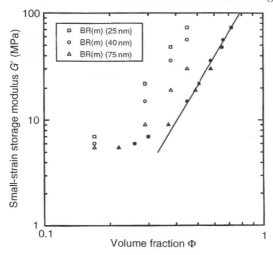

Fig. 10.14. Double logarithmic plot of the small-strain storage modulus versus filler volume fraction for E-SBR–BR(m) microgel composites with varying size of the BR microgels as indicated (open symbols). The solid line represents a master curve of slope 3.5 estimated from (10.37) with adapted layer thickness $\Delta = 2$ nm (solid symbols). Experimental data are taken from [262].

modulus comes from the deformation of filler clusters, which are then no longer rigid, but flexible. Under this condition the concentration dependence of the elastic modulus is described by (10.35) or alternatively by (10.37) if the effect of immobilized rubber cannot be neglected. Figure 10.14 shows a double logarithmic plot of the small-strain storage modulus versus filler volume fraction (open symbols) of heavily crosslinked (4% dicumylperoxide), hard BR microgels of varying size in E-SBR. Obviously, G'_0 increases with decreasing particle size. This behavior can be related to the increased amount of immobilized rubber with increasing specific surface of the filler particles. A common master curve for all data points can be constructed by estimating an effective volume fraction given by the term in square brackets in (10.37). This master curve with slope 3.5 is found from the closed symbols that are obtained by considering an effective volume fraction according to (10.37) for a common layer thickness $\Delta = 2$ nm, independent of particle size d and concentration ϕ.

The estimated layer thickness, $\Delta = 2$ nm, that results from the condition of minimum deviation from the master curve shown in Fig. 10.14 appears reasonable. It corresponds to a few layers of polymer segments that are fixed at the surface of the microgel clusters like a hard glassy skin. It is this immobilized rubber layer that gives the filler network a high stability compared to colloid networks aggregated in low-viscosity liquids. In particular, as shown by Payne [10], this leads to a shift of the critical strain amplitude at which filler network breakdown occurs by more than one order of magnitude if the strain amplitude dependence of the storage modulus of carbon-black-filled BR and liquid paraffin are compared. Furthermore, this critical strain amplitude is strongly affected by the surface chemistry of the filler particles, which influences the interaction strength between the particle as well as the amount of immobilized rubber.

Summing up, the CCA model assumes that: (i) filler flocculation in a rubber matrix starts above an aggregation limit ϕ^+ which depends on the interparticle distance and the affinity of the filler to the polymer as well as the mean distance between successive chain entanglements in the polymer melt; (ii) a transition in the cluster growth takes place in the networking regime $\phi > \phi^*$, where a space-filling configuration of clusters is reached and a macroscopically connected structure is formed; (iii) above the networking threshold ϕ^* a transition of the storage mechanism of elastic energy from the entropy-elastic rubber matrix to the energy-elastic filler network occurs; and (iv) the concentration dependence of the small-strain modulus is governed by a power law with exponent 3.5, which reflects the structure of the fractal heterogeneities of the filler network.

10.3 Stress-softening and quasistatic stress–strain modeling – the Mullins effect

10.3.1 The dynamic flocculation model

In this section, a constitutive model of hyperelasticity and stress softening of filler reinforced polymer networks is presented that is based on the extended tube model of rubber elasticity [42, 43, 57, 60, 101, 102, 105] (see Section 5.4.2) and a micromechanical theory of filler cluster breakdown in strained elastomer composites [58, 111, 123, 138, 151, 261, 263–266]. We focus on the mechanical response of reinforcing fractal filler clusters in elastomer composites up to large strain, leading to a microstructure-based model of stress softening and filler-induced hysteresis. Finally, the developed dynamic flocculation model is adapted to stress–strain cycles of filler-reinforced elastomers.

Reinforcement of polymer networks by nano-structure fillers like carbon blacks or silica is assumed to be related to the presence of physically bonded, fractal filler clusters in the rubber matrix that exhibit a particular size distribution $\phi(\xi_\mu)$. With increasing strain on a virgin sample, a successive breakdown of the filler clusters takes place under the exposed stress of the bulk polymer matrix. This begins with the largest clusters and continues to a minimum cluster size $\xi_{min} = \xi(\varepsilon_{max})$ which is specified below (equation (10.47)). During the back-cycle, i.e., the relaxation of the sample, reaggregation of the filler particles takes place, but the damaged filler–filler bonds that reform after having being broken differ from the original ones. Since, in general, the original bonds are annealed by the heat treatment during vulcanization (see Section 10.1.1), the damaged bonds are significantly softer and more fragile than the original ones. This mechanism implies a pronounced stress softening of the prestrained samples, since hard rigid cluster units are replaced by soft ones. Furthermore, in subsequent stress–strain cycles the reaggregated filler clusters with soft bonds bend substantially in the stress field of the rubber, implying that a certain amount of energy is stored in the clusters, which is dissipated when the clusters break. This mechanism leads to a filler-induced viscoelastic contribution to the total stress that impacts the internal friction or hysteresis of filled polymer networks, significantly. It is important to note that this kind of viscoelastic response is also present in the limit of quasistatic deformations, where no explicit time dependence of the stress–strain cycles is considered.

According to this model, two micromechanical mechanisms of cyclically strained reinforced polymer networks are distinguished. In the first, stress softening is considered to be related to hydrodynamic reinforcement of the rubber matrix by a fraction of hard filler clusters with strong filler–filler bonds in the virgin annealed state. In the second, filler-induced hysteresis results from a cyclic breakdown and reaggregation of the residual fraction of the more fragile filler clusters with softer already damaged filler–filler bonds. The fraction of rigid filler clusters decreases with increasing prestrain, while the fraction of fragile filler clusters increases. This leads to a shift of the boundary size ξ_{min} with increasing prestrain. The decomposition into hard and soft filler cluster units is depicted in Fig. 10.15.

Accordingly, we assume that for quasistatic, cyclic deformations of filler-reinforced rubbers up to large strain the total free energy density consists of two contributions:

$$W\left(\varepsilon_\mu\right) = \left(1 - \Phi_{\text{eff}}\right) W_{\text{R}}\left(\varepsilon_\mu\right) + \Phi_{\text{eff}} W_{\text{A}}\left(\varepsilon_\mu\right) , \qquad (10.38)$$

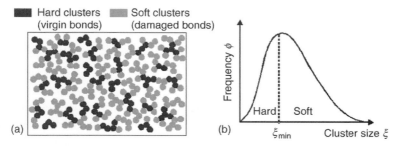

Fig. 10.15. (a) Schematic view of the decomposition of filler clusters in hard and soft units for preconditioned samples with virgin (annealed) and damaged filler–filler bonds. (b) The cluster size distribution with the prestrain-dependent boundary cluster size ξ_{min}. From [138].

where Φ_{eff} is the effective filler volume fraction. For well-dispersed spherical filler particles Φ_{eff} equals the filler volume fraction Φ. However, for structured particles like carbon blacks $\Phi_{\text{eff}} > \Phi$, depending on the grade number of the carbon black [138, 151]. The first addend of (10.38) takes into account the equilibrium energy density of the strained rubber matrix, including hydrodynamic reinforcement by the fraction of rigid filler clusters, as specified below. The second addend describes the energy stored in the substantially strained soft filler clusters:

$$W_A\left(\varepsilon_\mu\right) = \sum_\mu^{\dot{\varepsilon}_\mu<0} \frac{1}{2d} \int_{\xi_{\mu,\min}}^{\xi_\mu(\varepsilon_\mu)} G_A\left(\xi'_\mu\right) \varepsilon_{A,\mu}^2 \left(\xi'_\mu, \varepsilon_\mu\right) \phi\left(\xi'_\mu\right) \mathrm{d}\xi'_\mu . \qquad (10.39)$$

Here, d is the particle size and ξ_μ is the cluster size in spatial direction μ of the main axis system. $\phi\left(\xi_\mu\right)$ is the normalized size distribution of the clusters that is considered to be isotropic, i. e. $\phi\left(\xi_1\right) = \phi\left(\xi_2\right) = \phi\left(\xi_3\right)$. G_A is the elastic modulus and $\varepsilon_{A,\mu}$ is the strain of the soft filler clusters.

The dot in the upper limit of the sum in (10.39) denotes the time derivative, which means that the sum is taken over stretching directions with $\partial\varepsilon_\mu/\partial t > 0$ only. Consequently, clusters are strained and successively broken in stretching directions alone. Healing of the clusters takes place in the compression directions, implying that a cyclic breakdown and reaggregation of clusters can be described. The integration in (10.39) is performed over the fraction of soft filler clusters with a cluster size lying in the interval $\xi_{\mu,\min} < \xi_\mu < \xi_\mu\left(\varepsilon_\mu\right)$ that are not broken at exposed strain ε_μ of the actual cycle.

The clusters smaller than $\xi_{\mu,\min} = \xi_\mu\left(\varepsilon_{\mu,\max}\right)$, representing the fraction that survived the maximum exposed prestrain $\varepsilon_{\mu,\max}$ in a previous deformation cycle, are assumed to dominate the hydrodynamic reinforcement of the rubber matrix. Due to the stiff nature of their filler–filler bonds, corresponding to the bonds in the virgin state of the sample, these clusters can considered to behave quite rigidly. Accordingly, the contribution of the stiff clusters to the stored energy of the clusters W_A is neglected. Their mechanical action refers to an overstraining of the rubber matrix, which is quantified by a strain amplification factor X. This relates the external strain ε_μ of the sample to the internal strain ratio λ_μ of the rubber matrix:

$$\lambda_\mu = 1 + X\varepsilon_\mu . \qquad (10.40)$$

For the first deformation of virgin samples the strain amplification factor depends on the external strain $(X = X(\varepsilon_\mu))$. In the case of a prestrained sample and for strains smaller than the previous straining $(\varepsilon_\mu < \varepsilon_{\mu,\max})$, the strain amplification factor is constant and determined by $\varepsilon_{\mu,\max}$ $\left(X = X\left(\varepsilon_{\mu,\max}\right)\right)$.

In the following we apply an equation for the strain amplification factor of the overlapping fractal clusters (sufficiently high filler concentrations), as derived by

Huber and Vilgis [44,45] (see (8.26)). In the present case, the amplification factors $X\left(\varepsilon_\mu\right)$ and $X\left(\varepsilon_{\mu,\text{max}}\right)$ are evaluated by averaging over the size distribution of rigid clusters in all space directions. For prestrained samples this yields:

$$X\left(\varepsilon_{\mu,\text{max}}\right) = 1 + c\Phi_{\text{eff}}^{\frac{2}{3-d_{\text{f}}}}$$

$$\sum_{\mu=1}^{3}\frac{1}{d}\left\{\int_{0}^{\xi_{\mu,\text{min}}}\left(\frac{\xi_\mu'}{d}\right)^{d_{\text{w}}-d_{\text{f}}}\phi\left(\xi_\mu'\right)\mathrm{d}\xi_\mu' + \int_{\xi_{\mu,\text{min}}}^{\infty}\phi\left(\xi_\mu'\right)\mathrm{d}\xi_\mu'\right\}. \tag{10.41}$$

Here, c is a constant of order one, d_{f} is the fractal dimension and $d_{\text{w}} = 2d_f/D$ is the anomalous diffusion exponent on fractal clusters. For virgin samples, $X\left(\varepsilon_\mu\right)$ is obtained similarly by performing the integration in (10.41) from zero up to the strain-dependent cluster size $\xi_\mu\left(\varepsilon_\mu\right)$ (see (10.47)).

The elastic modulus G_{A} of the clusters entering (10.39) can be evaluated by referring to the Kantor–Webman model of flexible curved chain aggregates [257] (cf. Section 10.3.2). In a simplified approach introduced by Lin and Lee [37], the contributions from the two different kinds of angular deformation, bending and twisting, can be considered by an averaged bending–twisting deformation. This is obtained by replacing the elastic bending constant G through an averaged elastic constant $\overline{G}$. It yields in the case of CCA clusters (compare Sections 10.3.2)

$$G_{\text{A}} \equiv \xi^{-1}k_{\text{S}} = \frac{\kappa\overline{G}}{d^3}\left(\frac{d}{\xi}\right)^{3+d_{\text{f,B}}}, \tag{10.42}$$

where κ is a geometrical factor of order one and $d_{\text{f,B}}$ is the backbone fractal dimension of the filler clusters. For large clusters, the force constant of the cluster backbone $k \approx k_{\text{S}}$ is

$$k_{\text{S}} = \frac{\kappa\overline{G}}{d^2}\left(\frac{d}{\xi}\right)^{2+d_{\text{f,B}}}. \tag{10.43}$$

Equation (10.42) describes the modulus G_{A} of the clusters as a local elastic bending–twisting energy term $\overline{G}$ times a scaling function that involves the size and geometrical structure of the clusters.

To estimate the limiting cluster size $\xi_{\mu,\text{min}}$ as a function of external strain ε_μ, the properties of fractal filler clusters have to be described on a microscopic level. The failure or yield strain ε_{F} of the filler clusters results from the fact that a single cluster corresponds to a series of two molecular springs: a soft spring, representing the bending–twisting mode, and a stiff spring, representing the tension mode. The soft spring with force constant $k_{\text{S}} \sim \overline{G}$ impacts the elasticity of the whole system, since, in general, the deformation of the stiff spring can be neglected ((10.42) and

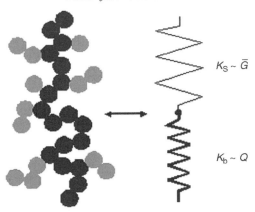

Fig. 10.16. Schematic view demonstrating the mechanical equivalence of a filler cluster and a series of soft and stiff molecular springs, representing the bending–twisting and tension deformation of filler–filler bonds, respectively. From [138].

(10.43)). The stiff spring governs the fracture behavior of the system, because it takes account of the longitudinal deformation and hence spatial separation of filler–filler bonds. Fracture of the cluster takes place, when a critical separation of bound filler particles is exceeded and the failure strain ε_b of filler–filler bonds is reached. The mechanical equivalence of a filler cluster and a series of two springs is illustrated in Fig. 10.16.

The failure strain ε_F of the filler cluster can be evaluated from the stress equilibrium between the two springs. With (10.43) one finds in the case of large clusters with $k_b \gg k_S$:

$$\varepsilon_F = \left(1 + \frac{k_b}{k_S}\right)\varepsilon_b \approx \frac{Q\varepsilon_b}{\kappa\overline{G}}\left(\frac{\xi}{d}\right)^{2+d_{f,B}}. \tag{10.44}$$

Here, Q is the elastic tension constant of the Kantor–Webman model and $k_b = Q/d^2$ is the force constant of longitudinal deformations of filler–filler bonds [257]. Equation (10.44) implies that the yield strain of a filler cluster increases with the cluster size ξ according to a power law. Furthermore, it is governed by the ratio of the elastic constants $Q/\overline{G}$. Consequently, larger clusters show a higher extensibility than smaller ones, due to the ability to bend and twist around the bonds and the clusters can survive up to large strain due to their high flexibility in strained rubbers. This kind of elastic behavior and the dependence of strength on cluster size plays a crucial role in stress softening and filler-induced hysteresis up to large strain.

Finally, to analyze the fracture behavior of filler clusters in strained polymer networks, we have to evaluate the strain $\varepsilon_{A,\mu}$ of the filler clusters relative to the external strain ε_μ of the sample. At medium and large strains, when there has been a stress-induced gel–sol transition of the through-going filler network (for $\varepsilon > 10\%$), the stress on the filler clusters is transmitted by the rubber matrix. Then, $\varepsilon_{A,\mu}$ follows from the stress equilibrium between the clusters and the rubber matrix $(\varepsilon_{A,\mu} G_A(\xi_\mu) = \widehat{\sigma}_{R,\mu}(\varepsilon_\mu))$. This yields with (10.42):

$$\varepsilon_{A,\mu}(\varepsilon_\mu) = \frac{d^3}{\kappa \overline{G}} \left(\frac{\xi_\mu}{d}\right)^{3+d_{f,B}} \widehat{\sigma}_{R,\mu}(\varepsilon_\mu) . \tag{10.45}$$

Here, $\widehat{\sigma}_{R,\mu}(\varepsilon_\mu)$ is the norm of the relative stress of the rubber matrix related to the stress at the beginning of each strain cycle, where $\partial \varepsilon_\mu / \partial t = 0$:

$$\widehat{\sigma}_{R,\mu}(\varepsilon_\mu) \equiv |\sigma_{R,\mu}(\varepsilon_\mu) - \sigma_{R,\mu}(\partial \varepsilon_\mu / \partial t = 0)| . \tag{10.46}$$

The application of this normalized, relative stress in the stress equilibrium equation (10.45) is essential for a constitutive formulation of cyclic cluster breakdown and reaggregation during every stress–strain cycle. It ensures that the clusters are not compressed but only stretched in spatial directions with $\partial \varepsilon_\mu / \partial t > 0$, since $\varepsilon_{A,\mu} \geq 0$ holds due to (10.46). In the compression directions with $\partial \varepsilon_\mu / \partial t < 0$ reaggregation of the filler particles takes place. By comparing (10.44) and (10.45) one finds that the strain $\varepsilon_{A,\mu}$ of the clusters increases faster with their size ξ_μ than the failure strain $\varepsilon_{F,\mu}$. Accordingly, with increasing strain the large clusters in the system break up first, followed by the smaller ones. The maximum size ξ_μ of clusters surviving at external strain ε_μ is found from the stress equilibrium between the rubber matrix and the failure stress $\sigma_{F,\mu}$ of the clusters $(\sigma_{F,\mu} = \varepsilon_{F,\mu} G_A(\xi_\mu))$:

$$\xi_\mu(\varepsilon_\mu) = \frac{Q \varepsilon_b}{d^2 \widehat{\sigma}_{R,\mu}(\varepsilon_\mu)} . \tag{10.47}$$

This allows an evaluation of the boundaries of the integrals in (10.39) and (10.41). Hence, the nominal stress contribution of the stretched filler clusters can be calculated. This is determined from $\sigma_{A,\mu} = \partial W_A / \partial \varepsilon_{A,\mu}$, where the sum over all stretching directions, which differ for the up and down cycles, has to be considered.

For uniaxial deformations $\varepsilon_1 = \varepsilon$, $\varepsilon_2 = \varepsilon_3 = (1 + \varepsilon)^{-1/2} - 1$, one obtains a positive contribution to the total nominal stress in the stretching direction for the up cycle:

$$\sigma_{0,1}^{up}(\varepsilon) = (1 - \Phi_{eff}) \sigma_{R,1}(\varepsilon) + \Phi_{eff} \widehat{\sigma}_{R,1}(\varepsilon) \int_{\frac{Q \varepsilon_b}{d^3 \widehat{\sigma}_{R,1}(\varepsilon_{max})}}^{\frac{Q \varepsilon_b}{d^3 \widehat{\sigma}_{R,1}(\varepsilon)}} \phi(x_1) \, dx_1 , \tag{10.48}$$

where

$$\widehat{\sigma}_{R,1}(\varepsilon) = |\sigma_{R,1}(\varepsilon) - \sigma_{R,1}(\varepsilon_{min})| \tag{10.49}$$

and $x_1 = \xi_1/d$. The stress of the rubber matrix $\sigma_{R,1} = \partial W_R/\partial \lambda_1$ can be evaluated with (5.42) by taking into account the strain amplification according to (10.40) and (10.41). For the down cycle in the stretching direction one finds a negative contribution to the total stress due to the norm in (10.46):

$$\sigma_{0,1}^{\text{down}}(\varepsilon) = (1 - \Phi_{\text{eff}}) \sigma_{R,1}(\varepsilon) - 2\Phi_{\text{eff}} \widetilde{\sigma}_{R,1}(\varepsilon) \int_{\frac{\varrho \varepsilon_b (1+\varepsilon_{\min})^{-3/2}}{2d^3 \widetilde{\sigma}_{R,1}(\varepsilon_{\min})}}^{\frac{\varrho \varepsilon_b (1+\varepsilon)^{-3/2}}{2d^3 \widetilde{\sigma}_{R,1}(\varepsilon)}} \phi(x_1) \, dx_1 .$$

$$(10.50)$$

The negative stress contribution results from the stretching of clusters in the lateral direction which hinders the relaxation of the polymer network during the back-cycle. In (10.50), the notation

$$\widetilde{\sigma}_{R,1}(\varepsilon) = \left| \sigma_{R,1}(\varepsilon) - \left(\frac{1 + \varepsilon_{\max}}{1 + \varepsilon} \right)^{3/2} \sigma_{R,1}(\varepsilon_{\max}) \right| \qquad (10.51)$$

has been used. The different choice of the extrema with $\partial \varepsilon_\mu/\partial t = 0$ in (10.49) and (10.51) is due to the fact that an up cycle begins at $\varepsilon = \varepsilon_{\min}$, but a down cycle begins at $\varepsilon = \varepsilon_{\max}$. As a rule, the relative stresses in the lower boundaries of the integrals in (10.48) and (10.50) have to be chosen in such a way that they attain their maximum values, implying that all soft clusters are broken and the condition $\xi_\mu = \xi_{\mu,\min}$ holds. Note that stress–strain cycles start or end at a strain between $\varepsilon_{\max}$ and $\varepsilon_{\min}$ cannot be described by the present model, since the reaggregation mechanism has not so far been considered.

Figure 10.17 shows an adaptation of the developed model for the up and down cycles ((10.48) and (10.50)) to experimental stress–strain data (equilibrium cycles) of a silica-filled ethlene-propene-diene rubber (EPDM) sample at different pre-strains. The extended tube model has been applied to model the rubber matrix (see Section 5.4.2). For the cluster size distribution the Smoluchowski approach for the kinetics of CCA of filler particles has been assumed with a fixed width parameter $\Omega = -0.5$ [138]:

$$\phi(\xi_\mu) = \frac{4d}{\xi_{0,\mu}} \left(\frac{\xi_\mu}{\xi_{0,\mu}} \right)^{-2\Omega} \exp \left(-\frac{(1 - 2\Omega) \xi_\mu}{\xi_{0,\mu}} \right) . \qquad (10.52)$$

Here $\xi_{0,\mu}$ is the mean cluster size, which is the same for all spatial directions $\mu = 1, 2, 3$ (isotropy assumption). Note that with this distribution function the integrals in (10.41), (10.49), and (10.50) can be solved analytically [266].

Figure 10.17 demonstrates that a good adaptation of the preconditioned samples can be obtained with a single set of polymer parameters $G_c = 1.09$ MPa,

$G_e = 0.6$ MPa and $n_e/T_e = 45$ (see Section 5.4.2), simply by varying the strain amplification factor X. The fitted parameters are $X_i = 7.520, 5.915, 4.733,$ $4.225,$ and 3.694 for increasing prestrains from $\varepsilon_{max} = 10\%, 20\%, 30\%, 40\%, 50\%,$ respectively. As discussed in Section 5.4.2, the tube constraint modulus G_e need not be treated as a fitting parameter but can be estimated from the plateau modulus $G_N \approx 1.2$ MPa ($G_e \approx G_N/2$). Further fitting parameters are the yield stress of damaged filler–filler bonds $Q\varepsilon_b/d^3 = 31$ MPa, the effective filler volume fraction $\Phi_{eff} = 0.26,$ and the relative mean cluster size $x_0 \equiv \xi_0/d = 10.1.$ Note that the value of Φ_{eff} is close to the filler volume fraction $\Phi \approx 0.23,$ indicating that the reinforcing silica particles are almost spherical.

The simulation data shown in Fig. 10.17 have been obtained without applying (10.41), since the strain amplification factor X is treated as an independent variable. However, it can be shown that the dependence of the fitting parameters X_i on ε_{max} is in good agreement with the predicted behavior of (10.41). This has been shown, e.g., for carbon-black- and silica-filled EPDM and SBR rubbers [266]. The dependence of the strain amplification factor $X_{max} \equiv X(\varepsilon_{max})$ on prestrain (equation (10.41), is depicted in Fig. 10.18 for various filled elastomer materials, as indicated. It corresponds to the well-known Payne effect of the storage modulus in the quasistatic limit, since X_{max} determines the initial slope of the hysteresis cycles, which can be compared with the storage modulus G' of non-linear viscoelastic materials. The

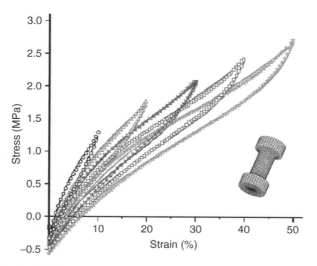

Fig. 10.17. Quasistatic stress–strain cycles in uniaxial extension (fifth cycles) of an EPDM sample with 60 phr silica coupled with silane (5 phr Si69) for prestrains ε_{max} between 10% and 50% (symbols) and adaptations using (10.48)–(10.52) (lines). Experimental data were obtained by using dumbbells as shown in the inset.

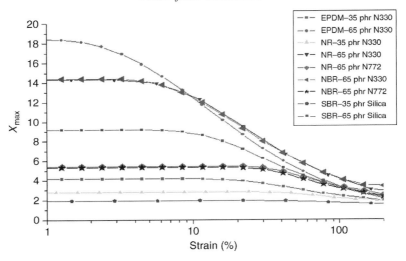

Fig. 10.18. Variation of the strain amplification factor $X_{max} \equiv X(\varepsilon_{max})$ with prestrain obtained from adaptations with (10.41) of uniaxial stress–strain cycles at prestrains between 1% and 100% of various filled rubber materials, as indicated.

data were obtained from adaptations of the fifth stress–strain cycle in the range 1–100 %, similarly to the ones shown in Fig. 10.17.

From the data in Fig. 10.18 it is obvious that in the large-strain regime X_{max} approaches a constant value scaling with the effective filler volume fraction Φ_{eff}, since the sum in (10.41) approaches 3 in the limit of large stress values $\sigma_{R,1}$. For small prestrains, the strain amplification factor levels out as well, since all clusters contribute to X_{max} in the limit of small stress values $\sigma_{R,1}$. Hence, the mean slope of the stress–strain cycles remains almost constant in the small-strain plateau regime, which is typically observed for the Payne effect in reinforced rubbers. Depending on the microstructure of the rubber and the type and amount of filler one observes characteristic differences that are also well known from dynamic-mechanical measurements of the Payne effect under harmonic excitations. Accordingly, the presented dynamic flocculation model provides a microstructure-based explanation of stress softening phenomena, also called the Mullins effect, as well as filler-induced hysteresis, which is found to be closely related to the Payne effect observed under dynamic loadings.

For a test of the developed model, we will next consider the prediction for the equi-biaxial deformation mode obtained with material parameters from uniaxial fits (plausibility test). Figure 10.19 compares simulations and experimental data of equi-biaxial stretching cycles between 20% and 80% prestrain for an S-SBR sample with 65 phr silica. For the smallest prestrains, agreement between experiment and simulation is fairly good, but with increasing prestrain one finds significantly more

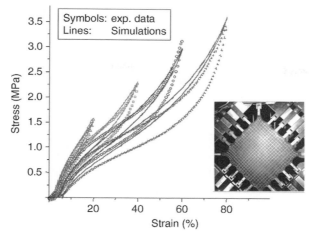

Fig. 10.19. Comparison of experimental data (symbols) and simulations (lines) of equi-biaxial hysteresis cycles between 20% and 80% prestrain for an S-SBR–silica sample with material parameters from fits to uniaxial stress–strain data. The inset shows the stretching frame used for equi-biaxial testing.

hysteresis for the experimental stress–strain curves than for the simulations. Nevertheless, the strain amplification factors, evaluated using (10.41), appear reasonable, since the simulated stress maxima fit quite well to the experimental data. To understand the large amount of hysteresis it is necessary to consider the equipment used for the equi-biaxial investigations more closely. The data were obtained with a stretching frame with 20 clamps holding the rubber sheet and rollers running on a steel frame, as depicted in the insert of Fig. 10.19. Due to this construction it is clear that the friction of the rollers, which increases with loading, contributes to the overall hysteresis measured between the up and down cycles. Accordingly, the successive deviations in hysteresis with increasing loading could be explained by the additional hysteresis resulting from the rollers of the stretching frame. A final answer to this question cannot be given at this stage, but further biaxial investigations, e.g. with the bubble inflation test, will have to be performed in the future.

A further question of interest is the temperature dependence of the hysteresis cycles. From the basic concepts of the dynamic flocculation model, it is clear that the main temperature effects on stress softening and filler-induced hysteresis result from the thermal activation of the virgin and damaged filler–filler bonds, respectively. Since these bonds are formed by flexible nanoscopic bridges of glassy polymer in the confined area between adjacent particles of the filler clusters, the thermal activation can be described by a simple Arrhenius dependence for the elastic constants Q_v and Q_d. This characteristic temperature dependence directly transfers to the two fitting parameters s_v and s_d, because the yield strains $\varepsilon_{v,b}$ and $\varepsilon_{d,b}$ of the virgin and damaged filler bonds can be considered to be temperature-independent parameters:

$$s_v \equiv \frac{Q_v \varepsilon_{v,b}}{d^3} = s_{v,\text{ref}} \exp\left[\frac{E_v}{R}\left(\frac{1}{T} - \frac{1}{T_{\text{ref}}}\right)\right], \tag{10.53}$$

$$s_d \equiv \frac{Q_d \varepsilon_{d,b}}{d^3} = s_{d,\text{ref}} \exp\left[\frac{E_d}{R}\left(\frac{1}{T} - \frac{1}{T_{\text{ref}}}\right)\right]. \tag{10.54}$$

Here, E_v and E_d are the activation energies of the virgin and damaged filler–filler bonds, respectively, R is the gas constant, and T temperature.

Figure 10.20 shows experimental data for uniaxial hysteresis cycles for EPDM–N339 samples at three different temperatures and compares them to the predictions of the model. The fifth up and down cycles are shown (symbols) together with simulations (lines) with material parameters obtained from fits at room temperature. For the activation energies the values $E_v = 3.62$ kJ/mol and $E_d = 4.91$ kJ/mol were used; these were not adapted to the stress–strain data but were derived from a master procedure for the frequency-dependent dynamic moduli G' and G'', respectively, at 3.5% strain amplitude [267]. The agreement between experimental data and simulations in Fig. 10.20 is fairly good.

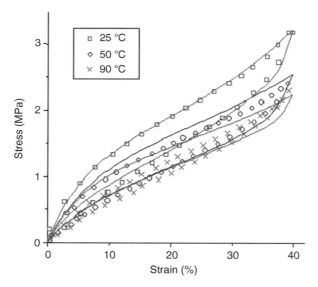

Fig. 10.20. Comparison of experimental data (symbols) and simulations (lines) of uniaxial hysteresis cycles for an EPDM sample with 60 phr N339 at 40% prestrain and three different temperatures, as indicated. Material parameters are taken from fits at room temperature ($T_{\text{ref}} = 25$ °C).

We finally note that the observed small permanent set of the experimental data in Fig. 10.17 has been taken into account by introducing suitable set-stresses, which are subtracted from the apparent stress $\sigma_{0,1}$ in (10.48) and (10.50). These set-stresses are found to be relatively small and can easily be extracted from the experimental data. Obviously, they have a negative sign and increase with rising prestrain. The equi-biaxial data in Fig. 10.19 cannot be precisely calibrated in the small-strain regime. Therefore, the simulation curves have been shifted slightly horizontally on the strain axis to compensate for experimental errors around the stress–strain origin. In Fig. 10.20 the experimental data have been shifted vertically on the stress axis to give better resolution of the results.

From the fair agreement between simulation curves and experimental stress–strain data for the different deformation modes it can be concluded that the extended tube model together with the dynamic flocculation model of cluster breakdown and reaggregation represents a good micromechanical basis for the description of stress softening and non-linear viscoelasticity of filler-reinforced elastomer materials. Thereby, the mechanisms of energy storage and dissipation are traced back to the elastic response of the polymer network as well as the elasticity and fracture properties of flexible filler clusters.

10.3.2 The Kantor–Webman model of flexible chain aggregates

The Kantor–Webman model represents a physical basis for various models of rubber reinforcement, which all are based on a fractal description of the filler clusters or networks (see Sections 10.2.4, 10.2.6, and 10.3.1). It describes the elasticity of curved elastic particle chains in a two-dimensional plane ($d = 2$) by referring to a vectorial Born-lattice model with a tension and bending energy term between neighboring bonds or particles. A chain is composed of a set of N singly connected bonds $\{b_i\}$ of length a under an applied force $\mathbf{F}$ at the two ends of the chain, as outlined in Fig. 10.21. The strain energy H is given by [257]:

$$H = \frac{F^2 N S_\perp^2}{2G} + \frac{a F^2 L_\|}{2Q},$$

(10.55)

where

$$S_\perp^2 = \frac{1}{F^2 N} \sum_{i=1}^{N} [(\mathbf{F} \times \mathbf{z})(\mathbf{R}_{i-1} - \mathbf{R}_N)]^2$$

(10.56)

is the squared radius of gyration of the projection of the chain on the two-dimensional plane and

$$L_\| = \frac{1}{a F^2} \sum_{i=1}^{N} (\mathbf{F} \bullet \mathbf{b_i})^2.$$

(10.57)

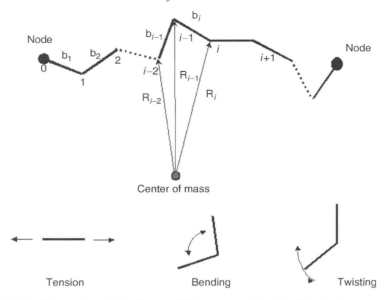

Fig. 10.21. Illustration of the Kantor–Webman model of flexible chain aggregates with tension, bending and twisting energy terms. From [138].

Here G and Q are local elastic constants corresponding to the changes of angles between singly connected bonds and longitudinal deformation of the bonds, respectively. The vector $\mathbf{z}$ is a unit vector perpendicular to the plane.

For long chains the second term in (10.55) can be neglected and most of the strain energy H results from the first bending term of the chain. Then, the force constant of the chain relating the elastic energy to the displacement squared of the end of the chain is given by:

$$k_S = \frac{G}{N S_\perp^2} . \tag{10.58}$$

In the three-dimensional case ($d = 3$), the angular deformation is not limited to in-plane bending, but also includes off-plane twisting. This makes the theory much more complex. To simplify the model, the contributions from these two angular deformations can be accumulated in the first term of (10.55) by replacing G with an averaged bending–twisting force constant $\overline{G}$ [37,268]. Then (10.58) is also valid for three-dimensional chain aggregates.

In applying (10.58) to fractal CCA clusters of bonded filler particles, one can use an approximation of the CCA cluster backbone as a single spanning arm, i. e. we describe it as a tender curved rod [151, 255] (compare Sections 10.2.6 and 10.3.1). This is possible because the CCA cluster backbone has almost no branches

[142, 259], implying that the energy of a strained cluster is primarily stored in filler–filler bonds along the connecting path between the backbone particles. Accordingly, the clusters act as molecular springs with end-to-end distance ξ, consisting of N_B backbone units of length d. The connectivity of the backbone units is characterized by the backbone fractal dimension $d_{f,B}$. Due to the fractal nature of CCA clusters,

$$N_B \cong \left(\frac{\xi}{d}\right)^{d_{f,B}}. \tag{10.59}$$

In the present approximation, $d_{f,B}$ is identified with the minimum (or chemical) fractal dimension, i. e. $d_{f,B} = d_{min} \approx 1.3$ for CCA clusters [142, 259]. Then, with $N = N_B$ and $S_\perp \cong \xi$, from (10.58) one obtains for the force constant k_S of the cluster backbone

$$k_S \cong \frac{\overline{G}}{d^2} \left(\frac{d}{\xi}\right)^{2+d_{f,B}}. \tag{10.60}$$

Finally, the elastic modulus of the cluster backbone is found as

$$G_A \equiv \xi^{-1} k_S \cong \frac{\overline{G}}{d^3} \left(\frac{d}{\xi}\right)^{3+d_{f,B}}. \tag{10.61}$$

Equation (10.61) describes the modulus G_A of the clusters via a local bending–twisting force constant $\overline{G}$ times a scaling function that involves the size and geometrical structure of the clusters. We point out that in the case of a linear cluster backbone with $d_{f,B} = 1$, (10.60) and (10.61) correspond to the well-known elastic behavior of linear flexible rods, where the bending modulus falls off with the fourth power of the length ξ. The above approach represents a generalization of this behavior to the case of flexible, curved rods.

The Kantor–Webman model, (10.55)–(10.58), can also be applied to percolation networks, or more precisely to LNB chains with characteristic rigid blobs [37, 268] (see Section 10.2.4). This is done simply by restricting the summation in (10.56) and (10.57) over the number L_1 of flexible, singly connected bonds. Then from simple scaling arguments one obtains a power-law dependence for the elastic modulus:

$$G_A \equiv \xi_p^{-1} k_S \sim (\Phi - \Phi_{crit})^\tau, \tag{10.62}$$

where Φ_{crit} is the critical filler volume fraction at the percolation threshold. The predicted exponent depends on the scaling exponents of percolation theory, and is $\tau \approx 3.6-3.7$ for the three-dimensional case ($d = 3$) [37, 257].

References

[1] S. Torquato, *Random Heterogeneous Materials* (Berlin: Springer, 2002).
[2] T. A. Vilgis and G. Heinrich, *Kautsch. Gummi Kunstst.* **48**, 323 (1995).
[3] J. Donnet, R. Bansal, and M. Wang (Eds.), *Carbon Black: Science and Technology* (New York: Marcel Dekker, 1993).
[4] J. Donnet and C. Lansinger, *Kautsch. Gummi Kunstst.* **45**, 459 (1992).
[5] A. Einstein, *Ann. Phys.* 289 (1906).
[6] L. Treloar, *The Physics of Rubber Elasticity*, 3rd edn (Oxford: Clarendon Press, 1975).
[7] S. Edwards and T. Vilgis, *Rep. Prog. Phys.* **52**, 247 (1988).
[8] G. Kraus (Ed.), *Reinforcement of Elastomers* (New York: Interscience, 1965).
[9] A. Payne, *J. Appl. Polym. Sci.* **6**, 57 (1962).
[10] A. Payne, *J. Appl. Polym. Sci.* **7**, 873 (1963).
[11] A. Payne, *J. Appl. Polym. Sci.* **8**, 2661 (1964).
[12] A. Payne, *Trans. IRI* **40**, T135 (1964).
[13] A. Payne, in *Reinforcment of Elastomers* (New York: Interscience, 1965), Chapter 3.
[14] A. Payne, *J. Appl. Polym. Sci.* **9**, 2273, 3245 (1965).
[15] A. Payne, *J. Appl. Polym. Sci.* **16**, 1191 (1972).
[16] A. Payne, *Rubber Chem. Technol.* **36**, 432 (1963).
[17] A. Medalia, *Rubber Chem. Technol.* **46**, 887 (1973).
[18] A. Medalia, *Rubber Chem. Technol.* **47**, 411 (1974).
[19] A. Voet and F. Cook, *Rubber Chem. Technol.* **40**, 1364 (1967).
[20] A. Voet and F. Cook, *Rubber Chem. Technol.* **41**, 1215 (1968).
[21] N. Dutta and D. Tripathy, *Kauschuk Gummi Kunstst.* **42**, 665 (1989).
[22] N. Dutta and D. Tripathy, *J. Appl. Polym. Sci.* **44**, 1635 (1992).
[23] N. Dutta and D. Tripathy, *Polymer Testing* **9**, 3 (1990).
[24] J. Ulmer, W. Hergenrother, and D. Lawson, *Rubber Chem. Technol.* **71**, 637 (1998).
[25] M.-J. Wang, W. Patterson, and G. Ouyang, *Kautschuk Gummi Kunstst.* **51**, 106 (1998).
[26] B. Freund and W. Niedermeier, *Kautschuk Gummi Kunstst.* **51**, 444 (1998).
[27] K. Mukhopadhyay and D. Tripathy, *J. Elastomers and Plastics* **24**, 203 (1992).
[28] M.-J. Wang, *Rubber Chem. Technol.* **71**, 520 (1998).
[29] A. Bischoff, M. Klüppel, and R. Schuster, *Polym. Bull.* **40**, 283 (1998).
[30] S. Vieweg, R. Unger, G. Heinrich, and E. Donth, *J. Appl. Polym.* **73**, 495 (1999).
[31] A. Payne and W. Watson, *Rubber Chem. Technol.* **36**, 147 (1963).
[32] T. Amari, K. Mesugi, and H. Suzuki, *Progr. Organic Coatings* **31**, 171 (1997).

[33] A. Payne and R. Wittaker, *Rheol. Acta* **9**, 91 (1970).

[34] A. Payne and R. Wittaker, *Rheol. Acta* **9**, 97 (1970).

[35] G. Kraus, *J. Appl. Polym. Sci.: Appl. Polym. Symp.* **39**, 75 (1984).

[36] A. van de Walle, C. Tricot, and M. Gerspacher, *Kautsch. Gummi Kunstst.* **49**, 173 (1996).

[37] C. Lin and Y. Lee, *Macromol. Theory Simul.* **5**, 1075 (1996).

[38] L. Mullins, *Rubber Chem Technol.* **21**, 281 (1948).

[39] L. Mullins and N. Tobin, *Rubber Chem. Technol.* **30**, 355 (1957).

[40] F. Bueche, in *Reinforcement of Elastomers* (New York: Interscience, 1965), Chapter 1.

[41] A. Medalia, *Rubber Chem. Technol.* **51**, 437 (1973).

[42] S. Edwards and T. Vilgis, *Polymer* **27**, 483 (1986).

[43] G. Heinrich, E. Straube, and G. Helmis, *Adv. Polym. Sci.* **85**, 33 (1988).

[44] G. Huber, *Universelle Eigenschaften gefüllter Elastomere*, Dissertation, Mainz University (1997).

[45] G. Huber and T. Vilgis, *Macromolecules* **35** (2002).

[46] E. Guth and O. Gold, *Phys. Rev.* **53**, 322 (1938).

[47] G. Heinrich and T. A. Vilgis, *Macromol.* **26**, 1109 (1993).

[48] U. Eisele and H.-K. Müller, *Kautschuk Gummi Kunstst.* **9** (1990).

[49] F. Bueche, *J. Appl. Polym. Sci.* **4**, 107 (1961).

[50] S. Govindjee and J. Simo, *J. Mech. Phys. Solids* **39**, 87 (1991).

[51] E. Dannenberg, *Rubber Chem. Technol.* **47**, 410 (1974).

[52] Z. Zigbi, *Adv. Polym. Sci.* **36**, 21 (1980).

[53] G. Hamed and S. Hatfield, *Rubber Chem. Technol.* **62**, 143 (1989).

[54] J. Haarwood, L. Mullins, and A. Payne, *J. Appl. Polym. Sci.* **9**, 3011 (1965).

[55] J. Haarwood and A. Payne, *J. Appl. Polym. Sci.* **10**, 1203 (1966).

[56] T. Witten, M. Rubinstein, and R. H. Colby, *J. Physique* **3**, 367 (1993).

[57] M. Klüppel and J. Schramm, in *Constitutive Models for Rubber I*, A. Dorfmann and A. Muhr (Eds.) (Rotterdam: A. A. Balkema Publishers, 1999), p. 211.

[58] M. Klüppel and J. Schramm, *Macromol. Theory Simul.* **9**, 742 (2000).

[59] M. Kaliske and G. Heinrich, *Rubber Chem. Technol.* **72**, 602 (1999).

[60] G. Heinrich and M. Kaliske, *Comput. Theoret. Polym. Sci.* **7**, 227 (1997).

[61] J. P. Flory, *Statistical Mechanics of Chain Molecules* (New York: Interscience, 1969); *Prog. Colloid Polym Sci.* **90**, 47 (1992).

[62] M. Doi and S. Edwards, *The Theory of Polymer Dynamics* (Oxford: Clarendon Press, 1986).

[63] P. de Gennes, *Scaling Concepts in Polymer Physics* (New York: Cornell University Press, 1979).

[64] R. P. Feynman and A. R. Hibbs, *Quantum Mechanics and Path Integrals* (New York: McGraw-Hill, 1965).

[65] R. Zallen, *The Physics of Amorphous Solids* (New York: Wiley, 1983).

[66] J. des Cloizeaux and G. Jannink, *Polymers in Solution: Their Modelling and Structure* (Oxford: Clarendon, 1990).

[67] C. Kittel, *Introduction to Solid Physics*, 7th edn. (New York: Wiley, 1996).

[68] J. P. Hansen and I. MacDonald, *Theory of Simple Liquids* (New York: Academic Press, 1986).

[69] M. Fixman, *J. Chem. Phys.* **23**, 1656 (1955).

[70] J. Zinn-Justin, *Quantum Field Theory and Critical Phenomena*, 2nd edn. (Oxford: Clarendon Press, 1993).

[71] J. Cardy, *Scaling and Renormalization in Statistical Physics* (Cambridge: Cambridge University Press, 1996).

[72] S. F. Edwards, *Proc. Phys. Soc.* **85**, 613 (1965).

[73] S. Edwards and P. Anderson, *J. Phys. F: Metal Phys.* **74**, 965 (1975).

[74] P. J. Flory, *The Principles of Polymer Chemistry* (Ithaca: Cornell University Press, 1953).

[75] W. H. Stockmayer, *J. Chem. Phys.* **12**, 125 (1944).

[76] D. Stauffer, *Introduction to Percolation Theory* (London: Taylor and Francis, 1985).

[77] J. P. Flory, *Proc. R. Soc.* **A 351**, 351 (1976).

[78] A. J. Stavermann, *Adv. Polym. Sci.* **25**, 609 (1982).

[79] P. de Gennes, *J. Phys. Lett. (France)* **37**, L1 (1976).

[80] F. Wall and J. Flory, *J. Chem. Phys.* **19**, 1435 (1951).

[81] R. C. Ball, M. Doi, S. F. Edwards, and M. Warner, *Polymer* **22**, 1010 (1981).

[82] W. Kuhn and F. Grün, *Koll. Z.* **101**, 248 (1942).

[83] S. Edwards and K. Evans, *J. Chem. Soc. Farad, Trans.* **2**, 145 (1981).

[84] S. Edwards, A. Chrompf, and S. Newman, *Polymer Networks* (New York: Plenum Press, 1971).

[85] R. Needs and S. F. Edwards, *Macromolecules* **16**, 1492 (1983).

[86] E. Helfand and D. S. Pearson, *J. Polym. Sci. , Polym. Phys. Symp.* **73**, 71 (1985).

[87] S. Edwards, *Br. Polym. J.* **9**, 140 (1977).

[88] M. Gottlieb and R. J. Gaylord, *Polymer* **24**, 1644 (1983).

[89] M. Gottlieb and R. J. Gaylord, *Macromolecules* **17**, 2024 (1984).

[90] M. Gottlieb and R. Gaylord, *Macromolecules* **20**, 130 (1987).

[91] R. Gaylord, *Polym. Bull.* **9**, 181 (1989).

[92] W. Graessley, *Adv. Polym. Sci.* **47**, 57 (1982).

[93] G. Marrucci, *Macromolecules* **14**, 434 (1981).

[94] H. E. Mark and B. Erman (Eds.) *Elastomeric Polymer Networks* (Englewood Cliffs: Prentice Hall, 1992).

[95] N. R. Langley, *Macromolecules* **1**, 348 (1968).

[96] K. Uyarama, T. Kawamura, and S. Kohjiya, *Macromoleules* **34**, 8261 (2001).

[97] G. Heinrich and E. Straube, *Acta Polymerica* **35**, 115 (1984).

[98] G. Heinrich and E. Straube, *Polym. Bull.* **17**, 247 (1987).

[99] M. Rubinstein and S. Panyukow, *Macromolecules* **30**, 8036 (1997).

[100] E. Straube, V. Urban, W. Pyckhout-Hintzen, D. Richter, and G. J. Glinka, *Phys. Rev. Lett.* **74**, 4464 (1995).

[101] M. Klüppel and G. Heinrich, *Macromolecules* **27**, 3569 (1994).

[102] M. Klüppel, H. Menge, H. Schmidt, H. Schneider, and R. H. Schuster, *Macromolecules* **34**, 8107 (2001).

[103] S. Kästner, *Faserforsch. & Textiltechnik* **27**, 1 (1976).

[104] S. Kästner, *Colloid Polym.* Sci. **259**, 499, 508 (1981).

[105] M. Klüppel, *Macromolecules* **27**, 7179 (1994).

[106] L. Mullins, *Appl. Polym. Sci.* **2**, 1 (1956).

[107] J. Scanlan, *J. Polym. Sci.* **43**, 501 (1960).

[108] G. Heinrich and W. Beckert, *Prog. Colloid Polym. Sci.* **90**, 47 (1992).

[109] M. Klüppel, *Prog. Colloid Polym. Sci.* **90**, 137 (1992).

[110] M. Klüppel, *J. Appl. Polym. Sci.* **48**, 1137 (1993).

[111] M. Klüppel, K. H. Luo and H. Schneider, *Macromolecules* **37**, 8000 (2004).

[112] L. J. Fetters, D. J. Lohse, D. Richter, T. A. Wilten, and A. Zirkel, *Macromolecules* **27**, 4639 (1994).

[113] M. Cates, *J. Phys. France* **46**, 1059 (1985).

[114] T. Vilgis, *J. Phys. II France* **2**, 1961 (1992).

[115] T. Vilgis, *J. Phys. II France* **2**, 2097 (1992).

[116] T. Vilgis, *Physica A* **153**, 341 (1988).

[117] D. Nelson, T. Piran, and S. Weinberg, *Statistical Mechanics of Membranes and Surfaces* (Singapore: World Scientific, 1989), Vol. 5.

[118] M. Daoud and P. de Gennes, *J. Physique* **38**, 85 (1977).

[119] C. Gay, P. G. de Gennes, E. Raphaël, and F. Brochard-Wyart, *Macromolecules* **29**, 8379 (1996).

[120] E. Raphael and P. Pincus, *J. Phys. II* **2**, 1341 (1992).

[121] F. Brochard and P. de Gennes, *J. Phys. (France)* **40**, L399 (1979).

[122] M. Daoud and J. Cotton, *J. Phys. France* **40**, 531 (1982).

[123] G. Heinrich and M. Klüppel, *Adv. Polym. Sci.* **160**, 2 (2002).

[124] G. Heinrich, L. Grave, and M. Stanzel, *VDI Berichte (German)* 49 (1995).

[125] L. Evans and W. Waddell, *Kautschuk, Gummi, Kunstst.* **48**, 718 (1995).

[126] R. Rauline, in *Rubber Compound and Tires Based on Such a Compound* (Paris: Michelin & Cie, 1992), p. 1.

[127] L. Reuvekamp, Reactive mixing of silica and rubber for tyres and engine mounts, Ph.D. thesis Twente University (2003).

[128] H. Palmgren, *Rubber Chem. Technol.* **48**, 462 (1975).

[129] H. Peitgen, H. Jürgens, and D. Saupe, *Chaos. Bausteine der Ordnung* (Berlin: Springer Verlag, 1994).

[130] P. van Buskirk, S. Turetzky, and P. Grunberg, *Rubber Chemistry and Technology* 577 (1975).

[131] S. Smale, *Bull. Am. Math. Soc.* 747 (1967).

[132] B. Mandelbrot, *Fractals: Form, Chance and Dimension* (New York: W. H. Freeman and Co., 1977).

[133] B. B. Mandelbrot, *The Fractal Geometry of the Nature* (New York: W. H. Freeman and Co., 1982).

[134] C. Herd, G. McDonald, and W. Hess, *Rubber Chem. Technol.* **65**, 107 (1992).

[135] M. Gerspacher and C. P. O'Farrell, *Kautsch. Gummi Kunstst.* **45**, 97 (1992).

[136] A. L. Mehaute, M. Gerspacher, and C. Tricot, in *Carbon Black Science and Technology*, J.-B. Donnet, R. C. Bansal, and M.-J. Wang (Eds.) (New York: Marcel Dekker, 1993).

[137] P. Meakin, B. Donn, and G. Mulholland, *Langmuir* **5**, 510 (1989).

[138] M. Klüppel, *Adv. Polym. Sci.* **164**, 1 (2003).

[139] C. Megaridis and R. Dobbins, *Combust. Sci. Tech.* **71**, 95 (1990).

[140] R. Samson, G. Mulholland, and J. Gentry, *Langmuir* **3**, 272 (1987).

[141] R. Ball and R. Jullien, *J. Phys. (Paris)* **45**, L103 (1984).

[142] P. Meakin, *Adv. Colloid Interface Sci.* **28**, 249 (1988).

[143] D. Sutherland, *Nature* **226**, 1241 (1970).

[144] A. Medalia, *J. Colloid Interface Sci.* **24**, 393 (1967).

[145] A. Medalia and F. Heckman, *Carbon* **7**, 567 (1969).

[146] R. Viswanathan and M. Heaney, *Phys. Rev. Lett.* **75**, 4433 (1995).

[147] F. Ehrburger-Dolle and M. Tence, *Carbon* **28**, 448 (1990).

[148] T. Rieker, S. Misono, and F. Ehrburger-Dolle, *Langmuir* **15**, 914 (1999).

[149] T. Rieker, M. Hindermann-Bischoff, and F. Ehrburger-Dolle, *Langmuir* **16**, 5588 (2000).

[150] J. Fröhlich, S. Kreitmeier, and D. Göritz, *Kautschuk Gummi Kunstst.* **51**, 370 (1998).

[151] M. Klüppel and G. Heinrich, *Rubber Chem. Technol.* **68**, 623 (1995).

[152] A. Schröder, *Charakterisierung verschiedener Rußtypen durch systematische statische Gasadsorption*, Ph.D. thesis, University of Hannover, 2000.

[153] A. Schröder, M. Klüppel, and R. Schuster, *Kautschuk Gummi Kunstst.* **52**, 814 (1999).

[154] A. Schröder, M. Klüppel, and R. Schuster, *Kautschuk Gummi Kunstst.* **53**, 257 (2000).

[155] R. Hjelm, W. Wampler, and P. Serger, in ACS Rubber Division Meeting, 21–24 May 1991 (American Chemical Society, 1991).

[156] A. Schröder, M. Klüppel, R. Schuster, and J. Heidberg, *Carbon* **40**, 207 (2002).

[157] M. Klüppel, A. Schröder, R. H. Schuster, and J. Schramm, in ACS Rubber Division Meeting, Dallas, Texas, 4–6 April 2000 (American Chemical Society, 2000).

[158] M. Klüppel, A. Schröder, G. Heinrich, R. H. Schuster, and J. Heidberg, in *Proceedings: Kautschuk-Herbst-Kolloquium 2000*, Hannover, p. 193 (unpublished).

[159] T. Zerda, H. Yang, and M. Gerspacher, *Rubber Chem. Technol.* **65**, 130 (1992).

[160] J. Isamail and P. Pfeifer, *Langmuir* **10**, 1532 (1994).

[161] J. Isamail, *Langmuir* **8**, 360 (1992).

[162] P. Pfeifer and M. Cole, *New J. Chem.* **14**, 221 (1990).

[163] A. Schröder, M. Klüppel, and R. Schuster, *Macromol. Mater. Eng.* **292**, 885 (2007).

[164] P. Pfeifer, M. Obert, and M. Cole, *Proc. R. Soc. Lond.* **A 423**, 169 (1989).

[165] B. Lippens, B. Linsen, and J. de Boer, *J. Catalysis* 32 (1964).

[166] G. Heinrich and M. Klüppel, *Kautschuk Gummi Kunstst.* **54**, 159 (2001).

[167] A. Adamson and J. Ling, *Adv. Chem.* **33**, 51 (1961).

[168] I. Langmuir, *J. Am. Chem. Soc.* **40**, 1316 (1918).

[169] R. Fowler and E. Guggenheim, *Statistical Thermodynamics* (Cambridge: Cambridge University Press, 1952).

[170] A. Schröder, M. Klüppel, R. Schuster, and J. Heidberg, *Kautschuk Gummi Kunstst.* **54**, 260 (2001).

[171] M. Douglas, *Macromolecules* **22**, 3707 (1989).

[172] T. A. Vilgis and G. Heinrich, *Macromolecules* **27**, 7846 .

[173] H. Luginsland, *Kautschuk Gummi Kunststoffe* **53**, 10 (2000).

[174] H. Luginsland, R. Krafczyk, and W. Lotz, in *Sulfur-Containing Organopolysiloxanes*, H.-D. Luginsland, R. Krafczyk, and W. Lotz (Eds.) (Germany: Degussa, 2000), p. 1.

[175] L. Reuvekamp, J. ten Brinke, P. van Swaaij, and J. Noordermeer, *Kautschuk Gummi Kunstst.* **55**, 41 (2002).

[176] L. Reuvekamp, J. ten Brinke, P. van Swaaij, and J. Noordermeer, *Rubber Chem. Technol.* 187 (2002).

[177] A. Guillet, J. Persello, and J. Morawski, Interaction of silica particles in a model rubber system: investigation by rheology and small angle scattering, ACS Rubber Division Meeting, Cincinatti, Ohio, October, 2000 (American Chemical Society, 2000), paper no 58.

[178] L. Ladouce, Y. Bomal, L. Flanckn, and D. Labarre, Dynamic mechanical properties of precipitated silica filled rubber: influence of morphology and coupling agent, ACS Rubber Division Meeting, Dallas, Texas, 4–6 April 2000 (American Chemical Society, 2000), paper no. 33.

[179] G. Huber and T. Vilgis, *Kautschuk Gummi Kunstst.* **52**, 102 (1999).

[180] H. M. Smallwood, *J. Appl. Phys.* **15**, 758 (1944).

[181] M. Klüppel and G. Heinrich, *Rubber Chem. Technol.* **68**, 623 (1995).

[182] A. I. Medalia, *Rubber Chem. Technol.* **51**, 437 (1978).

[183] A. Bunde and S. Havlin (Eds.), *Fractals and Disordered Systems* (Berlin: Springer Verlag, 1991).

[184] M. S. Jhon, R. J. Metz, and K. F. Freed, *J. Stat. Phys.* **52**, 1325 (1988).

[185] M. Cates and S. Edwards, *Proc. R. Soc.* **A 395**, 89 (1984).

[186] K. F. Freed and S. F. Edwards, *J. Chem. Phys.* **61**, 1189, 3626 (1974).

[187] D. Nelson, T. Piran, and S. W. (Eds.), *Statistical Mechanics of Membranes and Surfaces* (Singapore: World Scientific, 1989).

[188] T. Vilgis, *Phys. Rep.* **336**, 167 (2000).

[189] B. Meissner and L. Matejka, *Polymer* **43**, 3803 (2002).

[190] J. Sweeney, *Comp. Theor. Polymer Sci.* **9**, 27 (1999).

[191] C. Buckley and D. Jones, *Polymer* 3301 (1995).

[192] L. Landau and E. M. Lifschitz, *Elastizitätstheorie* (Berlin: Akademie-Verlag, 1983, 1986).

[193] P. Haronska and T. A. Vilgis, J. Chem. Phys. **102**, 6586 (1995).

[194] B. U. Felderhof and P. L. Iske, *Phys. Rev. A* **45**, 611 (1992).

[195] R. B. Jones and R. Schmitz, *Physica A* **122**, 105 (1983).

[196] R. M. Christensen, *Mechanics of Composite Materials* (Malabar: Krieger, 1991).

[197] M. Klüppel and G. Heinrich, *Adv. Pol. Sci.* **160**, 1 (2002).

[198] G. Huber, T. Vilgis, and G. Heinrich, *J. Phys. Condensed Matter* **8**, L409 (1996).

[199] M. Gerspacher, C. Farrell, L. Nikiel, and H. H. Yang, *Rubber Chem. Technol.* **69**, 569 (1996).

[200] M. Cates and R. Ball, *J. Phys. France* **49**, 2009 (1988).

[201] J. Feder, *Fractals* (New York: Plenum Press, 1988).

[202] J. Klein, in *Liquids at Interfaces, Les Houches 1988 Summer School XLVIII*, J. Charvolin, J. Joanny, and J. Z.-J. (Eds.) (Amsterdam: North Holland, 1990).

[203] D. Hone, H. Ji, and P. Pincus, *Macromolecules* **20**, 2543 (1987).

[204] A. Baumgärtner and M. Muthukumar, *Adv. Chem. Phys.* **XCIV**, 625 (1996).

[205] T. Garel and H. Orland, *Phys. Rev. B* **55**, 226 (1997).

[206] E. Eisenriegler, *Polymers near Surfaces* (Singapore: World Scientific, 1993).

[207] G. Huber and T. Vilgis, *Eur. Phys. J. B* **3**, 217 (1998).

[208] J. Machta, *Phys. Rev.* **A40**, 1720 (1989).

[209] J. Machta and T. Kirkpatrick, *Phys. Rev. A* **40**, 5345 (1990).

[210] V. G. Rostiashvili, M. Rehkopf, and T. A. Vilgis, *J. Chem. Phys* **110**, 639 (1999).

[211] V. G. Rostiashvili and T. A. Vilgis, *Phys. Rev.* **E62**, 1560 (2000).

[212] M. Rehkopf, V. G. Rostiashvili, and T. A. Vilgis, *J. Phys. (France) II* **7**, 1469 (1997).

[213] V. G. Rostiashvili, N. K. Lee, and T. A. Vilgis, *J. Chem. Phys.* **110**, 937 (2003).

[214] J. Bouchaud, *J. Phys (France)* **2**, 1705 (1992).

[215] G. Migliorini, V. Rostiashvili, and T. Vilgis, *Eur. Phys. J.* **33**, 61 (2003).

[216] A. Baumgärtner and M. Muthukumar, *J. Chem. Phys.* **94**, 4062 (1991).

[217] U. Ebert and L. Schäfer, *Europhys. Lett.* **21**, 741 (1993).

[218] W. Hess, *Macromolecules* **21**, 2620 (1988).

[219] K. S. Schweizer, *J. Chem. Phys.* **91**, 5802 (1989).

[220] W. Götze and L. Sjörgen, *Rep. Prog. Phys.* **55**, 241 (1990).

[221] G. Georg, A. Böhm, and N. Nguyen, *J. Appl. Polym. Sci.* **55**, 1041 (1995).

[222] J. Fröhlich and W. Niedermeier, in Proceedings: IRC 2000 Nürnberg, 4–7 September 2000, p. 107 (unpublished).

[223] J. Fröhlich and H. Luginsland, in Proceedings: Kautschuk-Herbst-Kolloquium 2000 Hannover, 6–8 November 2000, p. 13 (unpublished).

[224] T. Wang, M.-J. Wang, J. Shell, and N. Tokita, *Kautschuk Gummi Kunstst.* **53**, 497 (2000).

[225] J. Meier and M. Klüppel, *Macromol. Mat. Eng.* **293**, 12 (2008).

[226] R. Schuster, *Verstärkung von Elastomeren durch Ruß, Teil 1* (Wirtschaftsver band der Deutschen Kautschuk technologie, 1989), Vol. 40.

[227] G. Cotten, *Rubber Chem. Technol.* **48**, 548 (1975).

[228] G. Kraus and J. Gruver, *Rubber Chem. Technol.* **41**, 1256 (1975).

[229] B. Meissner, *J. Appl. Polym. Sci.* **18**, 2483 (1974).

[230] J. Brennan, T. Jermyn, and B. Boonstra, *Gummi Asbest Kunstst.* **18**, 266 (1965).

[231] X. Wang and C. Robertson, *Phys. Rev. E* **72**, 031406 (2005).

[232] C. Robertson and X. Wang, *Phys. Rev. Lett.* **95**, 075703 (2005).

[233] G. Heinrich, F. R. Costa, M. Abdel-Goad *et al.*, *Kautschuk Gummi Kunstst.* **58**, 163 (2005).

[234] G. Heinrich and M. Klüppel, *Kautschuk Gummi Kunstst.* **57**, 452 (2004).

[235] P. Leveresse, D. Feke, and I. Manas-Zloczower, *Polymer* **39**, 3919 (1998).

[236] M. Peschel and W. Mende, *The Predator–Prey Model: Do We Live in a Volterra World?* (Berlin: Akademie-Verlag, 1986).

[237] A. Kovacs, *J. Polym. Sci.* **30**, 131 (1958).

[238] G. McKenna, in *Comprehensive Polymer Science*, C. Booth and C. Price (Eds.) (Oxford: Pergamon, 1989), Vol. 12, Chapter 10.

[239] E. Nowak, J. Knight, E. Ben-Naim, H. Jaegen, and S. R. Nagel, *Phys. Rev. E* **57**, 1971 (1998).

[240] A. Barrat and V. Loreto, *Europhys. Lett.* **53**, 297 (2001).

[241] G. D'Anna, P. Mayor, G. Gremaud, A. Barrat, and V. Loveto, *Europhys. Lett.* **61**, 60 (2003).

[242] V. Trappe, V. Prasad, L. Cipellelt, P. N. Segre, and D. A. Weitz, *Nature (London)* **411**, 772 (2001).

[243] A. Liu and S. Nagel, *Jamming and Rheology: Constrained Dynamics on Microscopic and Macroscopic Scales* (New York: Taylor and Francis, 2001).

[244] A. Liu and S. Nagel, *Nature (London)* **396**, 21 (1998).

[245] A. Doolittle and D. Doolittle, *J. Appl. Phys.* **28**, 901 (1957).

[246] M. Fuchs and M. Cates, *Phys. Rev. Lett.* **89**, 248304 (2002).

[247] C. Robertson and X. Wang, *Europhys. Lett.* **76**, 278 (2006).

[248] A. R. Payne, in *Reinforcement of Elastomers*, G. Kraus (Ed.) (New York: Interscience, 1965), Chapter 3.

[249] S. Vieweg, R. Unger, K. Schröter, E. Donth, and G. Heinrich, *Polym. Networks Blends* **5**, 199 (1995).

[250] T. A. Vilgis and G. Heinrich, *Macromol. Symp.* **93**, 252 (1995).

[251] S. Vieweg, R. Unger, and E. Donth, *Kautschuk-Herbst-Kolloquium des DIK, Hannover 24–26,10.* (Hannover: DIK, 1996).

[252] J. Ulmer, *Rubber Chem. Technol.* **69**, 15 (1995).

[253] N. W. Tschoegl, *The Phenomenological Theory of Linear Viscoelasticity* (Berlin: Springer Verlag, 1989).

[254] M. Klüppel, R. H. Schuster, and G. Heinrich, ACS-Meeting, Rubber Div. Montreal May 5-8 (1996).

[255] T. Witten, M. Rubinstein, and R. Colby, *J. Phys. II* **3**, 367 (1993).

[256] D. Stauffer and A. Aharoni, *Introduction to Percolation Theory* (London: Taylor and Francis, 1995).

[257] Y. Kantor and I. Webman, *Phys. Rev. Lett.* **52**, 1891 (1984).

[258] P. G. Maier and D. Göritz, *Kautschuk Gummi Kunstst.* **49**, 18 (1996).

[259] P. Meakin, *Prog. Solid State Chem.* **20**, 135 (1990).

[260] A. Medalia, *J. Colloid Interface Sci.* **32**, 115 (1970).

[261] M. Klüppel, R. Schuster, and G. Heinrich, *Rubber Chem. Technol.* **70**, 243 (1997).

[262] S. Vieweg, R. Unger, E. Hempel, and E. Donth, *J. Non-Cryst. Solids* **235**, 470 (1998).

[263] M. Klüppel, J. Meier *et al.*, in *Constitutive Models for Rubber II*, D. Besdo, R. Schuster, and J. Jhlemann (Eds.) (Lisse: A. A. Balkema Publishers, 2001), p. 11.

[264] M. Klüppel, *Kautschuk Gummi Kunstst.* **50**, 282 (1997).

[265] M. Klüppel and G. Heinrich, *Kautschuk Gummi Kunstst.* **58**, 217 (2005).

[266] M. Klüppel, J. Meier, and M. Dämgen, in *Constitutive Models for Rubber IV*, P.-E. Austrell and L. K. (Eds.) (Lisse: A. A. Balkema Publishers, 2005), p. 171.

[267] A. L. Gal, X. Yang, and M. Klüppel, *J. Chem. Phys.* **123**, 014704 (2005).

[268] C. Lin and Y. Lee, *Macromol. Theory Simul.* **6**, 339 (1997).

[269] S. Brunauer, P. H. Emmer, and E. J. Teller, *J. Am. Chem. Soc.* **60**, 309 (1938).

[270] M. Kardar, G. Parisi, and Y. C. Zhang, *Phys. Rev. Lelt.* **56**, 889 (1986).

[271] A. Mildiev, V. G. Roshashvili, and T. A. Vilgis, *Europhys. Lett.* **61**, 340 (2004).

Index